624.154

THE COMMONWEALTH AND INTERNATIONAL LIBRARY
Joint Chairmen of the Honorary Editorial Advisory Board
SIR ROBERT ROBINSON, O.M., F.R.S. LONDON
DEAN ATHELSTAN SPILHAUS, MINNESOTA
Publisher: ROBERT MAXWELL, M.C., M.P.

STRUCTURES AND SOLID BODY MECHANICS DIVISION
General Editor: B. G. NEAL

THE DESIGN OF PILED FOUNDATIONS

T04312

THE DESIGN OF PILED FOUNDATIONS

by

THOMAS WHITAKER, D.SC., M.I.C.E.

PERGAMON PRESS
OXFORD . LONDON . EDINBURGH . NEW YORK
TORONTO . SYDNEY . PARIS . BRAUNSCHWEIG

Pergamon Press Ltd., Headington Hill Hall, Oxford
4 & 5 Fitzroy Square, London W.1
Pergamon Press (Scotland) Ltd., 2 & 3 Teviot Place, Edinburgh 1
Pergamon Press Inc., Maxwell House, Fairview Park, Elmsford, New York 10523
Pergamon of Canada Ltd., 207 Queen's Quay West, Toronto 1
Pergamon Press (Aust.) Pty. Ltd., 19a Boundary Street,
Rushcutters Bay, N.S.W. 2011, Australia
Pergamon Press S.A.R.L., 24 rue des Écoles, Paris 5ᵉ
Vieweg & Sohn GmbH, Burgplatz 1, Braunschweig

Copyright © 1970 Pergamon Press Limited

All Rights Reserved. No part of this publication may be reproduced, stored in a retrieval system, or transmitted, in any form or by any means, electronic, mechanical, photocopying, recording or otherwise; without the prior permission of Pergamon Press Ltd.

First edition 1970
Library of Congress Catalog Card No. 70–99995

Printed in Great Britain by W. & G. Baird Ltd., Belfast

This book is sold subject to the condition
that it shall not, by way of trade, be lent,
resold, hired out, or otherwise disposed
of without the publisher's consent,
in any form of binding or cover
other than that in which
it is published.

08 013952 3 (flexicover)
08 013953 1 (hard cover)

CONTENTS

	Page
ACKNOWLEDGEMENTS	ix

1. Introduction	1
Units of measurement	3

2 Where piles are used	5
The typical ground conditions	5
Site investigations	7

3 Types of piles and their construction	9
Pile classification	9
Displacement piles	9
Non-displacement piles	13
Screw piles	16
Pile installation	17
The structural strength of piles	22

4 Driving formulae	26
Introduction	26
Formula 1	27
Formula 2	27
Formula 3	28
Formula 4	30
Formula 5	31
Formula 6	31
Formula 7	31
Formula 8	32
Formula 9	33
Formula 10	33
The practical application of driving formulae	38
The limitations of driving formulae	42

5 Pile driving by vibration	47
Introduction	47
The theory of vibration driving	50
General comment	54
Vibratory impact drivers	54

6 The calculation of the ultimate bearing capacity of a pile from soil properties 56

Introduction 56
Early formulae 58
Recent formulae 61
Practical methods of calculating the ultimate bearing capacity 65
  Piles that are end-bearing in sand or gravel 66
    The standard penetration test 68
    The Dutch cone penetrometer 71
    Other types of penetrometers 77
  Piles in clay 78
    Load carried by the base of a pile in clay 78
    Driven piles 79
    Bored piles 82
    Large bored piles with enlarged bases 84
  Pile foundations carried to rock 91
    Driven piles 92
    Bored piles 93

7 The settlement of single piles and the choice of a factor of safety 96

Introduction 96
The choice of a factor of safety to ensure that the ultimate bearing capacity is not exceeded 97
The choice of a factor of safety to limit the settlement at working load 100

8 Piles in soft soils 109

The buckling of slender piles 109
Negative skin friction 112

9 Pile testing 119

Testing the condition of a pile after installation 119
Loading tests 121
  Loading systems 122
  The measurement of settlement 125
  Testing procedures 126
    The maintained load test 126
    The maintained load test for determining ultimate bearing capacity 128
    The constant rate of penetration test for determining ultimate bearing capacity 129

	CONTENTS	vii

10 Piles in groups with vertical loading — 135
Introduction — 135
Pile groups in clay — 137
　When not to use piles — 147
Pile groups in sand — 148

11 Horizontal forces on piles and pile groups — 152
Introduction — 152
Single piles — 154
Groups of vertical piles — 157
Groups with both vertical and inclined piles — 160
Graphical method for piles in three directions — 162
Methods based on elastic theory — 163
The choice of a design method — 164

12 The durability of piles — 166
Mechanical damage — 166
Chemical damage — 166
　Steel piles — 166
　Concrete piles — 169
Biological damage — 171
Damage due to industrial wastes — 172

REFERENCES — 173

INDEX — 179

ACKNOWLEDGEMENTS

I THANK the Director of the Building Research Station of the Ministry of Public Building and Works and H.M. Stationery Office for permission to reproduce information and figures previously published under Crown Copyright in papers of which I was the author or joint author. The figures are as follows: Figs. 6.14, 6.15, 6.16, 6.17, 7.1, 7.2, 7.3, 7.4, 9.5, 9.7, 9.8, 10.2, 10.3, 10.5, 10.6 and 10.7.

I also thank the following for permission to reproduce figures and other information:

Institution of Civil Engineers for figures appearing in Crown Copyright and other papers as follows: Figs. 6.15, 6.16, 6.17, 7.1, 7.2, 7.3, 7.4, 9.5, 9.7, 10.2 and 10.3, also for Table 6.1.

Institution of Civil Engineers and Prof. E. De Beer for Figs. 6.11 and 6.12.

British Steel Piling Co., Ltd. for Figs. 3.1(c), 3.6, 3.7, 3.8, 3.10, 3.12, 4.7 and 5.3 and for Tables 4.1 and 4.2.

John Wiley and Sons Ltd. for Figs 6.7 and 10.10.

Butterworths and Dr. C. van der Veen for Fig. 6.10.

American Society of Civil Engineers for Figs. 4.3 and 4.4.

Raymond International Concrete Pile Co. Ltd. for Fig. 3.1(a).

West's Piling and Construction Co. Ltd. for Fig. 3.1(b).

G.K.N. Foundations Ltd. for Figs. 3.2(c) and 5.4.

Prof. L. Zeevaert for the method of calculating skin friction.

The Public Works and Municipal Services Congress and Exhibition for Figs. 6.14, 10.5 and 10.6.

Peter Lind and Co. Ltd. for Fig. 3.12.

Frankipile Ltd. for Fig. 3.2(b).

Pressure Piling Co. (Parent) Ltd. for Fig. 3.11.

Central Electricity Generating Board, Southern Project Group and L. G. Mouchel and Partners for Fig. 3.9.

I also thank Mr. Saville Packshaw, Mr. D. J. Palmer and Mr. R. W. Cooke for reading the draft and for their helpful comments and Mr. G. L. Cairns formerly of B.R.S. and now with C.I.R.I.A. for translations from Russian literature. I am especially grateful to my wife for typing and checking the manuscript.

CHAPTER 1

INTRODUCTION

Until the beginning of the present century a "pile" was a straight log of timber about 300 mm (1 ft) in diameter and some 9 m (30 ft) or so long that was driven into the soil by the blows of a hammer. Piles or stakes that projected above the ground formed the supports for bridges and jetties and when driven entirely below the surface they were used to carry the walls and columns of buildings. Today, reinforced concrete and steel have largely taken the place of timber and although piles of these materials are driven like the timber log was driven, piling by another art has been developed. By making a tubular hole in the ground into which concrete is poured, a pile is formed when the concrete hardens.

Thus the pile of today has to be defined by its function rather than by its composition or mode of installation. Essentially a pile is an elongated or columnar body installed in the ground for the purpose of transmitting forces to the ground. This extremely wide definition covers piles formed in any manner and allows for the inclusion of piles with enlargements or protrusions at the sides or base.

When a pile carries a substantially axial force directed on to its head, as in the case of a vertical pile beneath a building, it is called a bearing pile. Piles are also used for resisting horizontal forces or moments, as in a jetty or dolphin. Where they are called upon to resist upward forces they may be called tension or anchor piles. "Sheet" piles are installed in rows and are shaped so that the sides of each pile interlock with those of its neighbours to form a continuous bulkhead, so providing a convenient method of constructing a cofferdam or retaining wall.

Piling is a form of construction of great antiquity, and an almost instinctive trust in piles for overcoming difficulties runs throughout foundation work. This attitude still exists today, fostered no doubt by the woeful lack of knowledge of how piles really behave and it must often have led to piles being installed where another type of foundation might have been preferable.

Although piling is still largely an art, and there are circumstances where dependence must be placed on experience or even on rule-of-thumb, the engineer endeavours as far as possible, to apply the methods of mechanics to piled foundation design. He aims to establish an adequate margin between the working load on the foundation and the load at which he estimates it will "fail" and also to ensure that the foundation does not settle more than can be tolerated under the working load.

This book deals with the theories which have been advanced to predict the loads which piles will carry, both singly and when used in groups to form a piled foundation. Thus, it is essentially about bearing piles, and when a pile is mentioned a bearing pile is implied unless otherwise stated. The problems of designing sheet piling are different, being concerned with the evaluation of soil pressures on the resulting bulkhead and its stability. They form another subject and will not be considered here.

It has been assumed that the reader has a working knowledge of the methods of site investigation and the testing of soil samples and is acquainted with the basic principles of soil mechanics. Certain theorems and analytical methods are common to the solution of all foundation problems and are best introduced to the student as part of the theory of surface foundations. For general information on these matters the student is referred to books such as *Soil Mechanics in Engineering Practice* by Terzaghi and Peck (second edition, 1967), *Fundamentals of Soil Mechanics* by Taylor (1948), *Foundation Design and Construction* by Tomlinson (1963).

Units of measurement

At the time of writing this book arrangements are being made for the United Kingdom to change to a metric system and, according to a programme produced by the British Standards Institution (1967), it is intended that the change shall be complete by the end of 1972. The system of units proposed is the Systeme International d'Unites (for which the abbreviation is SI) which takes as basic units the metre, kilogramme and second. The merit of the SI is that the derived units form a coherent system, that is, the product or quotient of any two unit quantities is the unit of the derived quantity. For example, unit mass multiplied by unit acceleration gives unit force; the unit of force being the Newton (N). Thus, the unit of stress or pressure becomes the Newton per square metre (N/m^2), If a submultiple is required this would be N/mm^2, while a multiple would be kN/m^2, since the preferred division of any unit is in multiples of 1000. The density of a material is expressed in kg/m^3, but the downward force generated by its weight is to be expressed in Newtons. The proposed units of weight and stress are thus profoundly different from the 'technical metric' units kgf and kgf/cm^2 at present used by civil and structural engineers in 'metric' countries and in Britain when metric calculations are involved.

The problem for the writer of a textbook is to present information in the way that the reader is likely to meet it in practice. The present indications are that the kilogramme force (kgf) rather than the Newton (N) will be used in civil and structural engineering when the British system of units is superseded, at least until engineers are satisfied that design processes will not be slowed down through lack of data expressed in Newtons. The adoption of the Newton may take some time, although the announcement by the Council of Engineering Institutions that only S1 units will be used in examinations from 1971 means that students will understand the use of the Newton. In consequence, this book is based principally on 'technical metric', but the alternative British and S1 units are given where appropriate.

Where in the particular context, however, the dimensions are empirically associated with one set of units these are used, the alternative being given if this seems appropriate.

CHAPTER 2

WHERE PILES ARE USED

The typical ground conditions

The purpose of any foundation is to transmit loads or forces to the ground without excessive settlement. A piled foundation is used where it is necessary to carry the load to an underlying stratum through a layer of weak or compressible material or through water. In a typical case the decision to use piling would probably be made if the site investigation showed a bed of rock, gravel or compact sand beneath deposits of alluvial silt, soft clay or peat which were too expensive to remove or to excavate through. Piles from 6 to 18 m (20 to 60 ft) long are commonly used in such circumstances; less frequent are cases where piles up to 36 m (120 ft) long are used and, exceptionally, piles may reach 60 m (200 ft) long.

It is important to differentiate between the various typical conditions in which piled foundations are employed. In Figs. 2.1 (a) and (b) the ground beneath the pile tips is strong for a very great depth. In Fig. 2.1 (c) the bearing stratum in which the piles terminate is of limited thickness and overlies weaker material. In a case of this sort the properties of the buried bed of weak material may predominantly influence the behaviour of the foundation.

The piles in Figs. 2.1 (a), (b) and (c) are called end-bearing or point-bearing piles, since most of the load they carry is transmitted to their lower ends or points, the shafts of the piles acting as columns.

Piling is often used in deep beds of clay as in Fig. 2.1 (d). The pile is supported in this case mainly by the adhesion or frictional

THE DESIGN OF PILED FOUNDATIONS

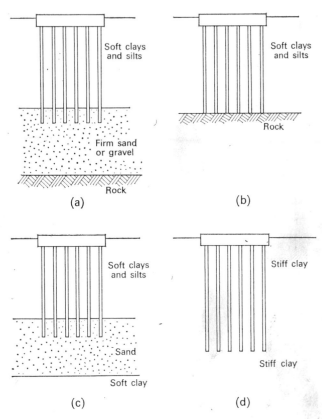

FIG. 2.1. The ground conditions in which piles are used. The piles in (a), (b) and (c) are end-bearing piles: in (d) they are friction piles.

action of the clay on the surface of the pile shaft. Such piles are termed friction piles.

All piles obtain support from both the frictional forces on the surface of their shafts and from direct bearing on their bases or points, but generally one of these components predominates and the division into "end-bearing" and "friction" piles is simply a convenient terminology.

When a pile is driven into loose sand compaction is caused by displacement and the vibration accompanying the installation. A group of piles sufficiently closely spaced will therefore create a dense block of sand a little larger than the volume bounded by the perimeter piles. This dense block with the embedded piles will transmit loads to its base and may meet the requirements of the design. Some other method of compacting the sand to form a dense block would serve equally well, and might be more economical than piling.

Site investigations

The primary requirements of a site investigation are that it should describe the ground conditions sufficiently well to enable a suitable bearing stratum to be chosen, and that it should extend sufficiently deeply below the likely level of the pile points to give information on all materials that might affect the foundation. This means that the soil should be examined to a distance of between 1 and $1\frac{1}{2}$ times the width of the structure below the pile points, unless there is definite evidence from other sources that no compressible materials are present. Many foundation failures have occurred because the soil was not investigated to an adequate depth. There is a strong temptation to terminate the investigation at, or a little below, the expected pile point level. Clearly with end-bearing piles, this is the level at which the principal loads will be applied to the soil, so that it is the strata below this level which need investigating.

The site investigation may be by boring only, or by boring accompanied by penetrometer tests such as the Standard Penetration Test or the Dutch Cone Penetrometer Test. (These are described later; see Chapter 6.) Unless the nature of the soil is known from previous trials it is advisable to put down two or more boreholes for soil identification and not to rely on the evidence of penetrometer tests alone. Soil samples from boreholes enable the types and positions of the different strata to be determined and if undisturbed samples are obtained, triaxial compression and oedometer tests may be made on cohesive soils. Where sands or gravels are encountered, penetrometer tests are generally the most satisfactory for giving a measure of the density or state of compaction of the bed. The contours of a resistant bed lying beneath weak material can be quickly found by probing to supply information for determining the pile lengths required. In soil containing boulders it is often difficult to differentiate between large boulders and beds of rock, so that caution is needed when interpreting the results if only one or two borings are put down. In such cases local knowledge, well drilling records and data from any geological survey made of the area will give guidance as to the nature of the strata likely to be encountered.

CHAPTER 3

TYPES OF PILES AND THEIR CONSTRUCTION

Pile classification

Piles may be classified in a number of ways, for example, by the material of which they are formed, or by their manner of installation. The method of classification used here is based on the effect the pile has on the soil during installation. This divides piles first into two families; those which displace the soil to accommodate the volume of the pile and those in which the soil is removed and the void formed is occupied by the pile. These are best called displacement and non-displacement piles respectively.

Displacement piles

Displacement piles are of two types:
(a) those in which a solid or hollow pile is driven into the ground and left in position;
(b) those in which a pile-like body is driven into the ground and then withdrawn, leaving a void which is filled with concrete during withdrawal.

It is convenient to speak of the pile being "driven" into the ground, but with piles of either type the actual method of installation might be by the blows of a hammer, by vibration, by pressure from a jack, by loading or by screwing, the choice being determined by the type of pile, the ground conditions and the circumstances of the situation.

Type (a) In piles of type (a) the essential feature is that the disturbed soil remains in contact with the surface of the newly introduced pile and there is an initial state of stress between the pile and the soil and in the surrounding soil resulting from the process of installation. Timber and solid precast concrete driven piles are obviously of this type, but so are the tubular and "shell" piles, provided the lower end is closed with a shoe to prevent the soil entering. A strong pipe or tube may be driven by hammer blows on the top and filled with concrete after installation. In shell piles the outer skin of the pile is first installed. This is in the form of a tube of thin sheet steel or lengths of concrete pipe, which in itself is too weak to withstand blows on the top without damage. One method of installation is to drive the shell by means of an internal steel mandrel onto which the shell is threaded. After the assembly has been driven into the soil to the required resistance, the mandrel is withdrawn, leaving the shell in position. The shell is then inspected internally and filled with concrete to complete the pile. With a steel shell pile the concrete core provides the principal structural strength. With a concrete shell pile some 50 per cent of the carrying capacity is supplied by the concrete shell, the concrete core providing the rest and making the pile into a monolith. The Raymond pile, Fig. 3.1(a), is a mandrel-driven steel-shell pile and the West pile, Fig. 3.1(b), is a mandrel-driven concrete-shell pile. Another method of installing steel shells is that used for the British Steel Piling Co.'s "cased pile", Fig. 3.1(c), in which a cushion of concrete is placed inside the shell on the plate which is welded on to close the base, and a drop hammer, working inside the shell, forces it down by blows on the concrete cushion. The pile is completed by filling the shell with concrete.

The advantages of steel pipe and steel and concrete shell piles lie in the ease with which the length of the pipe or shell can be altered to suit the ground conditions as driving proceeds, leading to economy over preformed piles.

Steel bearing piles of H section are useful on sites where a driven pile is required but the soil displacement must be kept to a minimum. The piles are usually driven by hammer, but a vibrator-driver can be used in suitable soils. Steel H piles are quickly extended by welding, or cut off as required.

TYPES OF PILES AND THEIR CONSTRUCTION

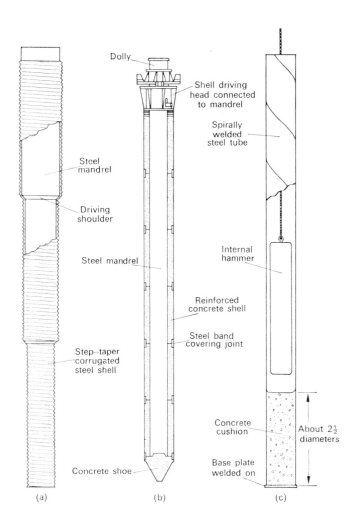

Fig. 3.1. Types of shell piles. (a) The Raymond step-taper pile, showing the driving mandrel. (b) The West pile, showing the driving mandrel and shell driving head. (c) The British Steel Piling Co.'s cased pile, driven by internal drop hammer.

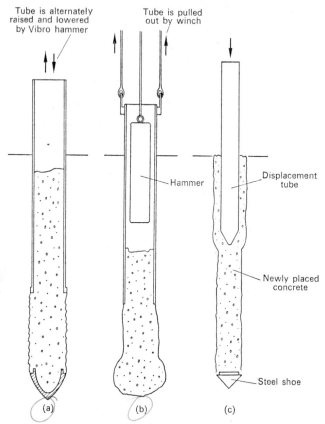

FIG. 3.2. Types of cast-in-place piles. (a) The Vibro pile. (b) The Franki pile (c) The Holmpress pile, showing the concrete and soil being compressed by the displacement tube.

Type (b). Piles of type (b) are generally formed by a proprietary system. In almost every case a strong tube, closed at the bottom by a temporary plug or a loose shoe, is first driven into the ground. Concrete is then placed in the tube and the tube is withdrawn; the plug or shoe is left behind and the concrete passes out at the now open bottom of the tube to fill the place previously occupied by the tube as the tube moves out of the way. The essential feature of these methods is that the concrete is cast-in-place in direct contact with the soil.

As the tube is withdrawn there will be some release of the stresses that were generated in the soil during driving and if the concrete is compressed as it is put into place restressing will result. In the Vibro system the tube is extracted by means of a hammer which gives it an up-and-down movement that tamps the outgoing concrete against the walls of the hole, as in Fig. 3.2(a). In the Franki system the tube is driven by an internal hammer falling onto a quantity of concrete placed in the bottom of the tube to form a plug. When the tube has been driven to the required depth the concrete plug is driven out, followed by the concrete to form the pile which is also driven out and compacted as the tube is raised. It is possible to form an enlarged base of compacted concrete by this method as shown in Fig. 3.2(b). In the Holmpress system the shaft is enlarged laterally and the soil compressed by driving another displacement tube within the newly placed plastic concrete and filling the void left on withdrawal with additional concrete, as shown in Fig. 3.2(c).

Non-displacement piles

Non-displacement piles are formed by removing the soil and putting the pile in its place. The many variants of this form of piling may be classified into two types:

(c) An open-ended tube is forced into the ground until it reaches the bearing stratum. The soil which enters the tube is removed and the tube is filled with concrete and left in position.

(d) A borehole is formed in the ground by a method appropriate to the particular type of soil. Concrete is cast-in-place in the hole, any temporary lining or casing to the sides of the hole being withdrawn so that the concrete is left directly in contact with the soil.

Type (c). Piles of type (c) are usually strong steel or reinforced concrete pipes, the lower end acting as a cutting edge, so that a core of soil enters the pipe. The soil is generally removed from the pipe as it is driven to prevent the formation of a plug, but when passing through soft strata where removal might cause inflow from the surrounding area, removal of the core may be delayed until the cutting edge has been sealed into ground which does not flow. The method admits of a number of variations in the way the soil is removed, e.g. by grab, by auger, by compressed air or water ejection, or by a separate inner liner to the main tube which is lifted out with the core of soil.

This type of pile possesses many of the advantages of the prefabricated displacement pile, with the additional merit that it causes little compression or displacement of the soil. These properties make it useful for installation near existing structures and for underpinning, in which case a pile may be built up by welding short lengths of pipe together as they are driven. The pipe is filled with concrete after cleaning out, this being essential if the pile is to function as an end bearing pile. The pipe itself usually provides a major proportion of the structural strength.

Type (d). The characteristic feature of piles of type (d) is that plastic concrete or mortar is cast-in-place against the soil in a borehole. The many systems differ mainly in the way the borehole is formed.

A borehole in clay up to about 600 mm (2 ft) in diameter may be formed with percussion tools on a rope handled by a simple tripod and winch, using casing tubes where necessary to prevent collapse of the sides of the hole. Holes up to 1.5 m (5 ft) diameter or larger may be taken out by mechanical augers, and this method is particularly economical in medium and stiff clays not requiring casing. A grab will operate in most types of soil; boulders or other obstructions too large for direct handling may be first broken down with a rock chisel.

Non-cohesive soils below the water table will flow into the borehole through the bottom of the casing due to water seepage, unless the casing is kept full of water to above the ground water level and the boring and excavation are carried out under water. By

TYPES OF PILES AND THEIR CONSTRUCTION

filling the borehole with a drilling mud the sides may be stabilised until it is convenient to enter the casing. In some forms of piling the drilling mud is relied upon to give the entire support and a casing is not used.

If the pile terminates in sand or gravel considerable skill is needed to ensure that the soil around the base is not loosened by inflowing water, as this would impair the bearing capacity of the pile. If the pile terminates on rock, provision may need to be made to drill the bottom of the casing tube into the rock, so that a seal is obtained, enabling the inside to be cleaned out and the rock surface inspected.

When placing the concrete in a bored pile care must be taken to ensure that it does not arch against the reinforcement or the sides of the hole thus forming a void. When a casing tube is used it must be withdrawn in such a way that the concrete runs out to fill the bore completely. To achieve this a good head of very workable concrete, of 130 to 150 mm (5 to 6 in.) slump, must be maintained in the casing. In poor work the sides of the hole may cave in, or the concrete may lift with the casing, forming a "waist" or even a complete gap in the shaft. Where the hole is filled with water or drilling mud, the concrete should be placed by means of a tremie pipe. Considerable skill, honest endeavour and attention to detail are needed to construct sound piles when casing is used, particularly in a water-filled hole. It should be borne in mind that

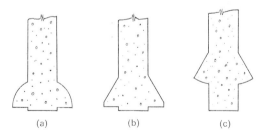

FIG. 3.3. Cross-sections of the enlarged bases produced by different types of under-reaming cutters.

the cost of finding out whether a pile is sound or not may be comparable with that of making another pile.

In stiff clays it is possible to enlarge the base of a bored pile either by hand digging or by an auger with blades that spread outwards to cut into the soil at the sides of the shaft to give a bell or conical shape as in Fig. 3.3. This process is often called under-reaming.

Screw piles

By providing a helical blade or a short length of screw thread at the base, a pile may be screwed into the ground. The shaft may be solid or hollow with a closed end, or the screw may be formed on a tube that is open at the bottom so that the soil enters and is removed, or the screw head may be detachable, concrete being placed through a withdrawable tubular shaft to form a cast-in-place pile shaft. The screw principle must be regarded,

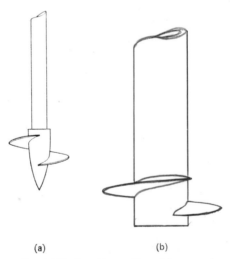

FIG. 3.4. Screw piles. (a) Early pattern with cast iron screw on a shaft about 250 mm (10 in.) diameter. (b) Modern screw-cylinder about 1 m (3 ft. 3ins.) diameter, open at the bottom, the core being removed as screwing proceeds.

therefore, as essentially a method of installation, since by this means piles of types (a), (b) or (c) can be formed. It is common practice for the diameter of the helical screw blade to be about 3 to 5 times that of the shaft and to be formed with 1 to $1\frac{1}{2}$ turns only, as shown in Fig. 3.4. Thus, the volume of soil displaced even with a "solid" shaft may be small in relation to the large bearing area of the blade. Modern designs have reinforced concrete tubular shafts up to about 1 m (3ft 3 in.) in diameter, with steel or reinforced concrete helices up to about 3 m (10 ft) diameter. Screw piles are valuable in maritime works since they can be made to resist upward as well as downward forces.

Pile installation

To describe in detail the various forms of piling equipment in use would be lengthy and out of keeping with the main purpose of this book. Each item of piling plant is usually built to suit a particular process and the ease and economy with which it will do its work is important to the specialist contractor in competitive tendering. Certain basic features will be mentioned, but for more complete descriptions the reader is referred to Chellis (1961),* Bachus (1961) and to the advertising literature of specialist contractors.

Drop hammers are widely used for driving piles, the hammer being lifted on a rope by a winch and allowed to fall onto the pile head by releasing a clutch on the winding drum, the falling hammer dragging the rope and reversing the motion of the drum. Sometimes, for making a driving test, provision is made for the hammer to be released from the rope by a trigger, allowing it to fall freely.

Power hammers are operated by steam, compressed air or internal combustion (diesel). Single acting steam and compressed air hammers are of two types; those in which the cylinder is held stationary, the hammer being lifted by the piston rod and then allowed to fall by gravity onto the pile head (Fig. 3.5) and those in which the piston is held and the cylinder itself forms the hammer

* References are given on pp. 173 ff.

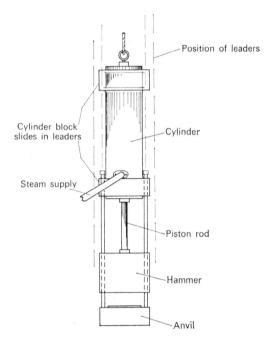

Fig. 3.5. Falling piston type of steam hammer.

(Fig. 3.6). With double acting hammers the ram is lifted and also driven downward by steam or compressed air. In the diesel hammer the head of the cylinder forms an anvil that rests on the head of the pile. The explosion in the cylinder lifts a heavy free piston and although at the same time an impulse is given to the pile, it is the piston falling back onto the anvil which imparts the main blow (Fig. 3.7). The length of the stroke varies slightly with the driving resistance, being greatest for hard driving.

Pile driving by vibration has been employed during the last two decades as an alternative to hammers. In this method the pile is vibrated in a vertical direction by a unit rigidly connected to the pile head. The vibration is communicated to the soil immediately

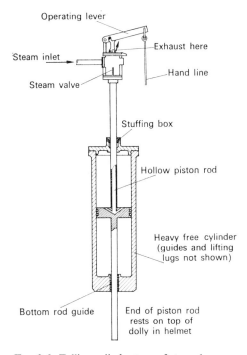

Fig. 3.6. Falling cylinder type of steam hammer.

around the pile, causing a reduction in shear strength and the pile sinks into the ground under its own weight and that of the vibrator unit. The method will be dealt with in Chapter 5.

An upright member or pair of members, called leaders are required with most piling systems of types (a), (b) and (c) to guide a drop hammer, power hammer or vibrator and to hold the pile or driven tube in alignment during the initial stages of pitching and driving. The leaders may be part of a framed structure supported on a base that carries the winches for handling the piles into place, operating the hammer and for lifting concreting skips and other equipment. The base is mounted on rollers or rail track

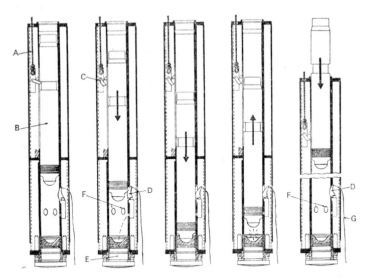

FIG. 3.7. The diesel hammer. (a) To start the hammer the piston B is lifted with the rope A until released by the automatic trip, C. (b) The falling piston actuates the fuel pump D and fuel is injected into the cup in the anvil E, the exhaust ports F are closed and compression occurs. (c) The piston strikes the anvil, splashing fuel into the compressed air. (d) The explosion drives the piston upward; exhaust occurs and air is taken in. (e) The piston falls and the cycle is repeated. The hammer is stopped by pulling the rope G.

as in Fig. 3.8. The whole unit is called a piling "frame". Frames of this type are commonly used in harbour and jetty work, but for land use there is demand for rigs with greater mobility. This is achieved in one modern design by mounting the leaders on a tracked chassis and supporting them by telescopic backstays, as shown Fig. 3.9. A more common practice is to hang the leaders from the top of the jib of a mobile crane, as in Fig. 3.8, the crane winches being used for the hammer and other work. Leaders that stand directly on the ground and are held upright by guy ropes are also used, usually with an attendant crane. The B.S.P. "cased pile" can be set in alignment in a simple timber trestle and the internal drop hammer operated by a crane.

Fig. 3.8. On the left, a piling frame mounted on a rail track; on the right, leaders hanging from a swivel at the top of the jib of a mobile crane.

Fig. 3.9. A B.S.P.-Priestman mobile self-erecting piling rig.

When a pile has to be driven through a bed of sand or sandy gravel, or the point has to be embedded some distance into such a bed, the risk of damaging the pile by prolonged heavy driving can be reduced by "jetting". In this operation, water is directed from nozzles at the point of the pile into the soil to loosen it and make driving easier. The jetting pipe may be forced down alongside the pile, or a water pipe with nozzles may be cast in a reinforced concrete pile when it is made. The former method is often preferred, since a clogged jet can be pulled up and cleared. Jetting must be stopped some distance before the pile point reaches its final position and the pile then driven on, so that the point is finally embedded in soil that has not been loosened by jetting. Jetting can be used with both hammer and vibrator driven piles. It is not effective in firm clays.

If driving is carried out incorrectly, breaking or crushing of the pile can occur. With timber piles, splitting or "brooming" at the head is obvious evidence of too hard driving, but this can also occur out of sight at the pile point if driving is continued after it has reached "refusal" in a resistant bed. Steel piles can be damaged at the head and if overdriven against an obstruction, the lower end may be crumpled or bent out of the straight.

To prevent damage of the head of a concrete pile by hammer blows a "helmet" is used, with a layer of packing, often called a

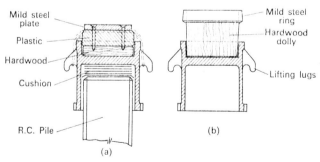

Fig. 3.10. Helmet for concrete piles, (a) with plastic dolly; (b) with timber dolly.

"cushion" in contact with the concrete. A "dolly" of timber or a plastic, such as nylon, protects the helmet itself. The arrangement is shown in Fig. 3.10. The common causes of failure of the concrete at the head of a pile are consolidation of the packing material and uneven placing of the packing on the pile head. Packing should be examined at regular intervals (say, every 1000 blows) and renewed if consolidated.

The choice of hammer weight depends on the plant in use, but with drop hammers and single acting hammers it is desirable that the weight of the hammer should be at least half that of the pile. With precast concrete piles it is preferable to use a hammer with a weight not less than 30 times the weight of 300 mm (1 ft) of pile in order to limit the driving stresses.

The variety of methods of forming boreholes for piles of type (d) has been mentioned earlier. Simple tripod rigs, as shown in Fig. 3.11, are commonly used for piles up to 600 mm (2 ft.) diameter. They are easily handled on weak ground or in low headroom. A large range of auger plant is available from the farm tractor mounting to the heavy crane mounted type shown in Fig. 3.12. In some types of plant, particularly those using grab equipment for piles up to 1 m (3 ft 3in.) diameter, special provision is made for working with casing tube. The tube is forced down and extracted by jacks and while it is in the ground it is continuously given a rotary oscillating motion to maintain it "free" in the soil.

The structural strength of piles

Any pile which is preformed must be strong enough to be put into place in the ground and all piles must be capable of withstanding the estimated forces brought onto them during service.

Most Codes of Practice dealing with piling contain recommendations for the quality of the materials and the details of manufacture of the different kinds of preformed piles. *Civil Engineering Code of Practice No.* 4 (1954), *Foundations*, covers timber, reinforced concrete and steel piles. Reference should be made to Saurin (1949) for current practice in the design of reinforced con-

FIG. 3.11. A tripod rig forming bored piles in a position with little headroom.

Fig. 3.12. A B.S.P. Calweld large diameter boring rig mounted on a crane.

crete piles and to Gardner and New (1961) and the Draft Standard Specification for pretensioned prestressed piles of the Prestressed Concrete Development Group (1964), for prestressed concrete piles.

In the case of precast reinforced concrete piles, with the exception of free standing piles in jetties, the amount of longitudinal reinforcement is determined by the bending moments caused in handing and pitching and this greatly exceeds what is needed once the pile is in the ground. It is common British practice to lift piles with a sling attached at two points and if these points are $L/5$ from the ends of the pile, as in Fig. 3.13, the bending moments will have the smallest value of $\pm WL/50$, where W is the pile weight and L the pile length. If the pile is pitched on end by a

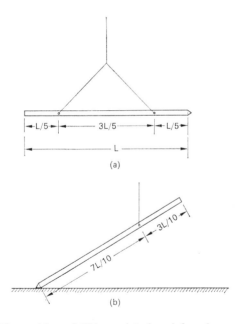

FIG. 3.13. The position of lifting points in reinforced concrete piles, (a) for lifting and stacking with a two point lift, (b) for pitching upright with a one point lift.

sling at one point, the bending moments have the smallest value $\pm WL/22$ when the point is $3L/10$ from the end. Saurin (1949) suggests that due to the possibility of mishandling, a value of $\pm WL/18$ should be allowed as the maximum moment, which allowing a factor of safety of $2\frac{1}{4}$ gives a moment of resistance for design purposes of $WL/8$. In American practice multiple point slinging is more common and the amount of longitudinal reinforcement used is smaller.

The lifting points should be predetermined by casting in eyes or tubes. With a square section pile a tube through which a pin can be passed enables a bridle to be used and this will minimise the risk of the pile being lifted with a diagonal of the section vertical, the resisting moment in the diagonal direction being inadequate. When piles are stacked for curing, care must be taken to place the supports beneath the lifting points.

Lateral reinforcement in the form of hoops is necessary in both precast R.C. and prestressed concrete piles. To reduce breakage by hammer at the pile head the volume of lateral reinforcement should be not less than 0.6 per cent of the gross volume for a length of about three times the pile width down the pile. Elsewhere it may be merely nominal, but should be not less than 0.2 per cent.

A major contribution to the understanding of the stresses in reinforced concrete piles during driving was that of Glanville, Grime, Fox and Davies (1938). The transmission of a wave in an elastic rod due to a blow at one end had been studied by St. Venant (1867) and the suggestion that pile driving was of a similar nature was first made by Isaacs (1931). Glanville *et al.* showed experimentally that the hammer impact sent a compression wave down the pile, which was reflected from the base as a compression or tension wave depending on whether the base was in contact with a strong support or not (cf. the reflection of sound waves in an organ pipe). Tension caused in this manner may occur in a pile when the point is in weak material, the packing is hard and the hammer rebounds. The most likely cause of tension cracks in a pile, however, apart from those caused during handling, are due to transverse bending, which can occur if there is a long length of pile out of the ground and the blow is struck eccentrically on the pile head.

Glanville *et al.* made recommendations for avoiding breakage in driving due to excessive compressive stresses. The most favourable driving conditions occur when the heaviest hammer is used, with packing or cushion material of the lowest stiffness on the pile head, the height of drop of the hammer being adjusted to give a safe stress in the pile.

Bored piles, being formed in place, are only called upon to resist the forces occurring during service. Since little is known about the bending moments actually brought onto a pile beneath a foundation, there is some divergence of opinion as to the amount of reinforcement required. In the main, the forces will be axial and in large diameter piles the reinforcement is often nominal and limited to the upper 6 m (20 ft).

CHAPTER 4

DRIVING FORMULAE

Introduction

It is natural for anyone driving a stake into the ground to assume that the effort needed depends on the "resistance" of the ground. For nearly two centuries engineers have applied this idea to pile driving and many mathematical expressions termed "driving formulae" or "dynamic formulae" have been devised for calculating the resistance. All driving formulae owe their existence to the assumption that the driving resistance is equal to the ultimate bearing capacity of the pile under static loading.

Driving formulae are based on the action of the hammer on the pile in the last stage of its embedment and it is assumed that this can be represented by some simple mechanical principle. No more factors than are necessary for the solution of the assumed principle are introduced and in no case is the train of events following the blow fully accounted for. Thus driving formulae are simple idealisations of a complex event; how remote from fact they may be will be seen in the following derivations of some of the more common formulae.

The symbols used in pile driving formulae are well established, and will be used in this chapter.

W = weight of the hammer;
P = weight of the pile;
H = height of fall of the hammer;
R = the driving resistance;
s = the set, i.e. the net distance the pile is driven by a blow;
A = cross-sectional area of the pile;
L = length of the pile;
E = Young's Modulus of the pile.

Formula 1

It is assumed that:
(a) the hammer and pile may be treated as impinging particles;
(b) the hammer gives up its entire energy on impact;
(c) on impact a resistance R to the motion of the pile is immediately generated which remains constant while the pile moves a distance s.

The available energy of the hammer is WH and the work done in overcoming the resistance is Rs, so that

$$WH = Rs \tag{4.1}$$

This is an elementary formula enabling R to be calculated from measured values of W, H and s. It formed the basis of the Sanders formula of 1851.

Formula 2

It is assumed that:
(a) and (b) are as in Formula 1;
(c) under the hammer blow the resistance increases to the value R in an elastic manner as the pile is displaced, remains constant for further displacement and then falls to zero in an elastic manner as the pile "rebounds".

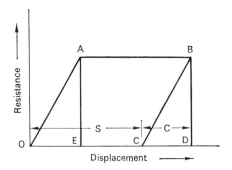

FIG. 4.1. Resistance–displacement diagram of the pile for one hammer blow.

The sequence of events is represented by the lines $OABC$ in the resistance-displacement diagram, Fig. 4.1. The maximum displacement is OD and the pile rebounds so that its final displacement is OC, i.e. $OC = s$. The work done against the resistance in reaching the displacement OD is represented by the area $OABD$. The area BDC represents elastic energy given up as the pile rebounds. The hammer has to supply energy to displace the pile the distance OD, but since the energy of rebound is dissipated, the energy usefully employed is given by area $OABC$. If it is assumed that the lines representing the initial elastic displacement and the rebound are parallel, then

Total work done $= OABD$

$$= OABC + BDC$$

Thus $\qquad WH = R(s + c/2) \qquad (4.2)$

where c is the elastic displacement of the pile head.

A formula of this type was published by A. M. Wellington in *Engineering News* in 1888 and is most usually called the *Engineering News* formula. In this formula empirical values in inches are given to the term $c/2$, so that for compatibility H and s are also to be measured in inches.

For drop hammers $\quad WH = R(s + 1.0)$
For single acting steam hammers
$$Wh = R(s + 0.1)$$

Formula 3

The assumptions are the same as for Formula 2.

Clearly if the impulse is not great enough to overcome the driving resistance the hammer will bounce and the pile will not be driven. Referring to the resistance-displacement diagram, Fig. 4.1, driving just commences when the hammer provides energy equivalent to area OAE. If the required height of fall for this is H_0, the hammer energy is WH_0.

But $$OAE = CBD = Rc/2$$

Therefore $$WH_0 = Rc/2$$

And from Eq. (4.2)
$$WH = Rs + WH_0 \tag{4.3}$$

H_0 is found in practice by obtaining the sets for different values of H, plotting the graph of H versus s and finding the intercept on the H axis of the line drawn through the points, as in Fig. 4.2.

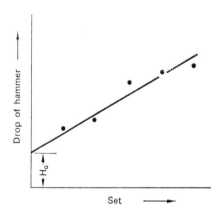

FIG. 4.2. Plot of hammer drop versus set, to determine H_0, the drop at which driving just begins.

In Morrison's formula of 1868 the sets s_1 and s_2 for two heights of drop H_1 and H_2 are found, thus:

$$WH_1 = Rs_1 + Rc/2$$
$$WH_2 = Rs_2 + Rc/2$$

and therefore
$$W(H_1 - H_2) = R(s_1 - s_2) \tag{4.4}$$

Formula 4

It is assumed that:

(a) the hammer and pile may be treated as impinging particles having a coefficient of restitution e and that Newton's laws of impact apply;

(b) an energy equation of the form

$$WH = Rs + U \qquad (4.5)$$

applies to the driving of a pile, where U is the energy supplied by the hammer not usefully absorbed in advancing the pile;

(c) the only energy lost is due to impact.

According to Newton, the loss of energy at impact between two bodies of masses M and m, having velocities V and v is

$$\frac{(1 - e^2)Mm(V - v)^2}{2(M + m)}$$

Putting $M = W/g$, $m = P/g$, $V = 2gH$ and $v = 0$, then

$$U = \frac{(1 - e^2)PWH}{(W + P)}$$

Inserting this in Eq. (4.5) gives

$$\frac{WH(W + e^2P)}{(W + P)} = Rs \qquad (4.6)$$

An allowance for loss of energy in this manner is made in a number of formulae.

If it is assumed that $e = 0$ then

$$\frac{W^2H}{(W + P)} = Rs \qquad (4.7)$$

This is the "Dutch", or Eytelwein's formula, published about 1812.

Formula 5

It is assumed that:
(a) the energy equation $WH = Rs + U$ may be applied;
(b) during the blow the pile is compressed elastically as if it were a strut under a static load R and the only energy lost is due to this elastic compression.

The elastic compression of the pile is RL/AE and the elastic energy is $R^2L/2AE$, so that $U = R^2L/2AE$.
Thus

$$WH = Rs + R^2L/2AE \tag{4.8}$$

This is Weisbach's formula of about 1850.

Formula 6

It is assumed that:
(a) there is frictional or other loss in the hammer system, so that the energy actually applied at impact is kWH, where k is a constant less than 1;
(b) there is loss due to elastic compression of the pile as in Formula 5;
(c) there is loss due to impact as in Formula 4 and this may be stated empirically.

Thus, according to Janbu (1953) who introduced empirically derived constants, the energy equation is

$$\frac{kWH}{(1 \cdot 5 + 0 \cdot 3P/W)} = \frac{R^2L}{2AE} + Rs \tag{4.9}$$

Formula 7

It is assumed that:
(a) there is frictional loss in the hammer system, the energy at impact being kWH;
(b) the elastic compression of the pile is that which would occur if all the available hammer energy were used in causing

the compression, i.e. the elastic compression is $(2kWHL/AE)^{\frac{1}{2}}$.

Thus
$$kWH = Rs + \frac{R}{2}(2kWHL/AE)^{\frac{1}{2}} \qquad (4.10)$$

This was termed the "Danish" formula by Sørensen and Hansen (1957).

Formula 8

It is assumed that there are losses of energy
(a) in the hammer system;
(b) due to impact;
(c) due to elastic compression of the pile;
(d) due to elastic compression of the head assembly comprising the dolly, helmet and packing;
(e) due to elastic compression of the ground.

Thus if L', A' and E' are the equivalent length, area and Young's Modulus of the head assembly and c_q is the compression or quake of the ground, it is assumed that the energy equation is

$$kWH = Rs + kWHP\frac{(1+e^2)}{(W+P)} + \frac{R^2L}{2AE} + \frac{R^2L'}{2A'E'} + \frac{Rc_q}{2} \qquad (4.11)$$

This is Redtenbacher's "complete" formula of 1859.

By putting $\dfrac{R^2L}{2AE} = c_p$ and $\dfrac{R^2L'}{2A'E'} = c_c$

substitution in Eq. (4.11) gives

$$\frac{k(W+e^2P)WH}{(W+P)} = R\left[s + \frac{1}{2}(c_c + c_p + c_q)\right] \qquad (4.12)$$

which is Hiley's formula of 1925. The more usual form of this formula is $R = WH\eta/(s + c/2)$, in which $\eta = k(W + e^2P)/(W + P)$ and $c = c_c + c_p + c_q$.

Formula 9

Formulae of purely empirical type have been devised from collections of data by engineers experienced in pile driving.

A formula of simple type is that due to Cornfield (1961, 1964), who found that an approximation to within $\pm$ 6 per cent of the result obtained by Hiley's driving formula for reinforced concrete piles was given by the expression

$$R = 0.08W(2 + H)(140 - L)(1.0 - s) \qquad (4.13)$$

where R = ultimate driving resistance in tonf
W = weight of hammer in tonf
H = drop of hammer in feet;
L = length of the pile in feet;
s = the final set in inches per blow.

The applicability of this formula is limited to reinforced concrete piles of length 20 to 80 ft, a hammer drop of 3 to 5 ft and a set of not more than 0·33 in. per blow, but preferably in the range 5 to 10 blows per inch. The constants apply for a single acting hammer, timber dolly and 2 to 3 in. of packing under the helmet. If a plastic or greenheart dolly is used R should be increased by about 10 per cent. If a winch operated drop hammer is used, the drop H should be multiplied by 0·9.

Formula 10

In all the above formulae, excepting the purely empirical, it is assumed that a force is generated instantly throughout the pile on impact, which is incorrect. The work of Glanville, Grime, Fox and Davies (1938) showed that pile driving was a phenomenon which depended on the transmission of compression waves down the pile. In their theoretical examination of the problem, they assumed the pile to behave like an elastic rod to which the wave equation

$$\frac{\partial^2 \xi}{\partial t^2} = \frac{E}{\varrho} \frac{\partial^2 \xi}{\partial x^2} \qquad (4.14)$$

would apply, where ξ is the displacement of an element at a distance x from the head of the pile at time t, E is the Young's Modulus and ϱ the density of the pile. The resistance was assumed to be entirely at the base of the pile and to be directly proportional to the base displacement; side friction was neglected and the cap and packing were assumed to behave as an elastic spring in order to enable an integration of the wave equation to be performed. The principal interest of Glanville *et al.* was in the stresses caused by the blow, but others have pointed out that the method could be made to serve as the basis of a driving formula.

Smith (1955, 1962) introduced a versatile method by which integration of the wave equation by a finite difference approach was obtained. To perform the calculation the pile is divided into a number of equal lengths. Each element of length is then represented by a weight joined to the adjacent weights by springs. The hammer and cushion are represented by a weight and spring at

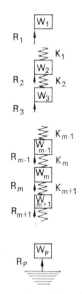

Fig. 4.3. A pile represented by weights and springs.

the head. Resistances are applied to the weights to represent the shaft friction and the base resistance. This system is shown in Fig. 4.3.

The basic equations for the motion and forces for the general case of weight W_m are

$$\left. \begin{array}{l} D_m = d_m + V_m \Delta t \\ C_m = D_m - D_{(m+1)} \\ F_m = C_m K_m \\ Z_m = F_{(m-1)} - F_m - R_m \\ V_m = v_m + Z_m g \Delta t / W_m \end{array} \right\} \qquad (4.15)$$

where
C = spring compression in time interval n
c = spring compression in time interval $(n-1)$
D = displacement in time interval n
d = displacement in time interval $(n-1)$
F = force exerted by spring in time interval n
K = spring constant
R = resistance in time interval n
Δt = time interval
V = velocity in time interval n
v = velocity in time interval $(n-1)$
W = weight
Z = accelerating force in time interval n
g = acceleration due to gravity

Having established the basic equations, Smith then introduced the soil properties termed "quake" and "viscosity" into the problem. Referring to the idealised resistance–displacement diagram for the pile element W_m given in Fig. 4.4, CD represents the rebound or quake, which Smith assumes to be the same for all elements of the pile and which is given the symbol Q. Assuming OA and CB are parallel, $OE = Q$ and the maximum value of the soil resistance on element W_m is equal to Q/K'_m, where K'_m is the equivalent "spring constant" of the soil. If the displacement of the soil is D'_m, the displacement of the pile element relative to the soil will be $(D_m - D'_m)$ and the resistance of the soil due to this displacement will be $K'_m (D_m - D'_m)$. The numerical value of $(D_m - D'_m)$ cannot exceed Q.

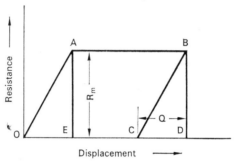

Fig. 4.4. The resistance–displacement diagram for a single element.

The viscous resistance of the soil to the motion of the pile element is assumed to be proportional to its velocity, so that in any instant the static resistance would be increased by a viscous resistance $J'v_m$, where J' is a viscous or "damping" constant applicable to the pile shaft.

The total resistance on the pile element is thus

$$R_m = K'_m(D_m - D'_m)(1 + J'v_m)$$

In the case of the point section, the constant J is used. The point resistance is thus

$$R_p = K'_p(D_p - D'_p)(1 + Jv_p)$$

Smith allowed for hysteresis in the compression of the packing on the pile head by a modification of the equation for F_1. With packing having a coefficient of restitution e

$$F_1 = (K_1 C_1/e^2) - [(1/e^2) - 1] K_1 C_{1\ max}$$

Where $C_{1\ max}$ is the maximum value of C_1.

Smith outlined a computer routine whereby the equations could be solved to give the "set" and the stresses in the pile for a given ultimate resistance. The various quantities entering the calculation are

1. The weight of the hammer.
2. The velocity of the hammer at impact.
3. The coefficient of restitution of the packing.
4. The weights and spring constants of the pile elements.
5. The time interval used in calculation.
6. The ultimate resistances contributed by the shaft and the base.
7. The ground quake and damping factors for the soil.

Equations (4.15) are solved for each of the pile elements involved (i.e. from $m = 1$ to $m = p$) for a succession of time intervals starting with $n = 0$ as the moment the first weight (i.e. the hammer) travelling with known velocity V_1 touches the first spring. Smith recommends that Δt be taken as 1/3000 or 1/4000 sec and the succession continued until D_p passes a maximum.

Forehand and Reese (1964), following Smith's suggestions, examined what values of J, Q and the proportions of the total resistance carried by the shaft and the base were needed to obtain correlation with test loadings made on piles. They used a series of pile tests for which the relevant data were available and took in their calculations the following ranges of values:

For sand $Q = 0.0042$ to 0.017 ft (1.27 to 5.08 mm)
$J = 0.10$ to 0.20 s/ft (0.328 to 0.656 s/m)
For clay $Q = 0.0042$ to 0.025 ft (1.27 to 7.62 mm)
$J = 0.40$ to 1.00 s/ft (1.31 to 3.28 s/m)

The value of J' was arbitrarily taken at $J/3$.

Their results, although encouraging, showed up the serious lack of knowledge of the essential parameters needed for this method of approach.

Smith was careful to point out that the resistance found by this method was that which existed at the moment of driving. To obtain the load-bearing capacity of the pile some time afterwards introduced the unknown effect of the recovery with time of the soil remoulded and disturbed during driving.

The practical application of driving formulae

It is generally assumed that a driving formula for a drop hammer may be applied to single acting steam and compressed air hammers, the weight and drop of the moving part being taken directly into the formula as W and H. Friction of the piston and back pressure of the exhaust cause loss of energy, which may be covered by an appropriate empirical value of k, so that the available energy per blow is kWH. Experience shows in the case of double-acting

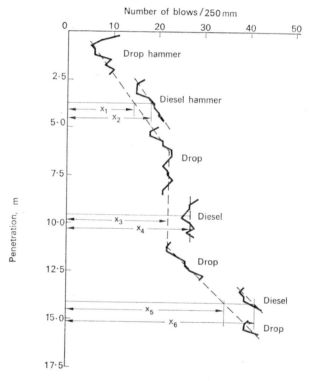

FIG. 4.5. Comparison of the driving records of diesel and drop hammers on the same pile. Three different beds are encountered, giving three comparison ratios x_1/x_2, x_3/x_4 and x_5/x_6.

DRIVING FORMULAE

hammers, which deliver blows in rapid succession, that the penetration per blow depends on the number of blows per minute. It is therefore essential that the manufacturers rated energy per blow at the speed of operation when taking the "set" is used in the driving formula in place of kWH.

In the case of diesel hammers, the manufacturer's rated energy per blow corresponding to the stroke of the hammer when taking the set may be used in place of kWH in a driving formula. When this information is not available, correlation with a drop hammer may be obtained by driving a pile partly by one hammer and then changing to the other and observing the pattern of the curve of blows per unit distance of penetration versus the pile penetration. In Fig. 4.5 the unit of distance for each blow count is 250 mm, and from this curve values of the ratio of the number of blows per 250 mm by drop hammer to the number of blows per 250 mm by diesel hammer in each soil bed may be found. Thus, using this ratio for other similar piles driven with the same diesel hammer on the same site, the equivalent drop-hammer blow-count can be calculated, enabling a driving formula intended for drop hammers to be used for the diesel hammer.

The movement of the pile head when struck by the hammer may be determined by means of the apparatus shown in Fig. 4.6(a). A piece of card is fixed to the pile at a convenient height and a straight edge is supported close to the card on stakes set about 1.20 m (4 ft) from the pile. By drawing a pencil steadily along the straight edge to make a line on the card as a series of blows is struck, a trace is obtained, as shown in Fig. 4.6(b) recording the set s and the sum of the temporary compression of the pile and the ground $(c_p + c_q)$. By this method the compression in the length of the pile above the card is unaccounted for; generally it is small enough to be neglected.

The temporary compression in the helmet assembly c_c cannot be determined by field measurements, but may be estimated by first assuming a value for the ultimate driving resistance and calculating the driving stress on the cross-sectional area of the pile

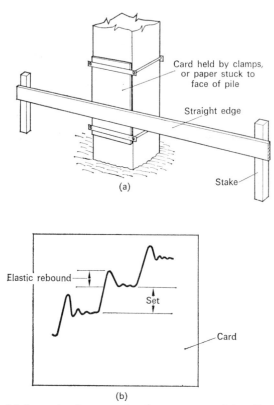

Fig. 4.6. (a) Apparatus for measuring the movement of the pile relative to the ground. A pencil is drawn steadily along the straight edge as the blows are struck. (b) Trace produced on the card for three successive blows.

and assuming that the packing and dolly, which are the compressible parts of the head assembly, are stressed to this value.

Their compressions may be obtained from Fig. 4.7, which is compiled from data given in *Civil Engineering Code of Practice No. 4, Foundations* and *B.S.P. Pocket Book*, issued by the British Steel Piling Co. Ltd. The compression of a complete head assembly consisting of dolly, helmet and packing is given by the

DRIVING FORMULAE

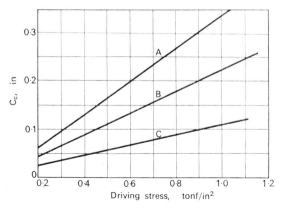

Fig. 4.7. Graph for estimating empirically the compression of the dolly and packing. A, 3 in. packing under the helmet on a concrete pile. B, Helmet with dolly on a concrete or steel pile. C, 1 in. pad only on a concrete pile.

TABLE 4.1

VALUES OF e

Type of pile	Head condition	Drop, single-acting or diesel hammers	Double-acting hammers
Reinforced concrete	Helmet with composite plastic or greenheart dolly and packing on top of pile	0·4	0·5
	Helmet with timber dolly,† and packing on top of pile	0·25	0·4
	Hammer direct on pile with pad only	—	0·5
Steel	Driving cap with standard plastic or greenheart dolly	0·5	0·5
	Driving cap with timber dolly†	0·3	0·3
	Hammer direct on pile	—	0·5
Timber	Hammer direct on pile	0·25	0·4

† Not greenheart.

sum of the ordinates to curves *A* and *B* at the calculated driving stress.

Empirical values of e, the coefficient of restitution, and of k, the hammer coefficient, are given in Tables 4.1 and 4.2, which have been taken from *B.S.P. Pocket Book*.

The value of H for the blows for which the set is taken should be measured on a suitable scale, and not estimated "by eye".

TABLE 4.2

HAMMER COEFFICIENT k

Hammer	k
Drop hammer operated by a trigger release	1·0
Drop hammer operated by releasing the winch clutch and overhauling the rope	0·8
Single-acting steam or compressed air hammer	0·9

The limitations of driving formulae

The assumptions and omissions made in creating Formulae 1 to 9 pay little regard to the actual forces and motions occurring during driving a real pile, or to the nature of the soil and its behaviour. For example, the assumption that the impact of a hammer on a pile is a problem to which Newton's laws may be applied is incorrect, since these laws apply only to particles capable of free motion. It is similarly an over-simplification to assume that the dynamic resistance can be appropriately expressed by a single force in a simple energy formula and the calculation of an energy correction for the elastic compression of the pile as if the loading were static is also unrealistic. Finally, there is no basis at all for assuming that the dynamic resistance is equal to the static load bearing capacity.

Although a formula based on the wave equation shows promise, since it meets many theoretical objections, its practical value is seriously limited by lack of precise information about quake, the viscous soil resistance, the distribution of resistance down the shaft and the proportions contributed by the shaft and the base to the total resistance. One obstacle to its use, of course, is that an electrical computer is required to produce an answer, so that the formula is at present largely of academic interest, although it has been used for checking practical pile driving problems.

Hiley's formula is used in Britain more than any other; variants of the *Engineering News* formula are most commonly used in the United States, and other countries have other preferences. Of the simple formulae, the one termed Formula 3 has a directness of approach which warrants greater attention than it has hitherto received.

No formula, however impressive, can overcome the errors inherent in the assumptions from which it is derived, yet although these assumptions will not bear rigorous examination, many engineers continue to use one or other of the formulae, because they are easy to apply under site conditions and give a result cheaply and quickly.

The acceptability of any formula can be best examined by comparing ultimate bearing capacities derived by it with those determined by loading tests. Terzaghi (1942) did this with data from tests on 39 timber, concrete and steel piles using seven different formulae. He found that the ratio μ of the real load to the computed load covered the range 0·25 to 4·0, that the range varied for different formulae applied to the same data and that the same formula was not necessarily "good" for timber, concrete and steel piles. Others have made similar collections of data and have assessed the relative merits of different formulae by statistical analysis. Sørensen and Hansen (1957) used a collection of 78 test results with the Eytelwein, Hiley, Janbu and Danish formulae and also a numberical integration of the wave equation. Agerschou (1962) used 171 test results with the *Engineering News*, Weisbach and Danish formulae. These authors found that log μ followed an approximately normal (Gaussian) distribution, and presented their results as probability graphs, with abscissae log μ and ordinates a normal probability scale. For each formula, at a given value of $\log \mu$ the ordinate represented the percentage of tests having $\log \mu$ values smaller than or equal to that value of $\log \mu$. From these graphs can be obtained the values of $\log \mu$ at which $2\frac{1}{2}\%$ of tests would give lower values of $\log \mu$, and also the values of $\log \mu$ at which $2\frac{1}{2}\%$ of tests would give higher values of $\log \mu$. The range of $\log \mu$ covered by the central 95% of tests is a measure of the reliability of the formula. The lower and upper limits and the range for the central 95% are given in Table 4.3 for the various formulae. The ranges show that the Eytelwein and *Engineering News* formulae have poor reliability compared with the others. There is nothing to choose between the Hiley, Janbu and Danish formulae, while the Weisbach formula is a little less reliable.

TABLE 4.3

Author	Formula	Value of log μ			Median value of μ	F	
		at lower $2\frac{1}{2}\%$ limit	at upper $2\frac{1}{2}\%$ limit	95% range		at lower $2\frac{1}{2}\%$ limit	at lower 5% limit
Sørensen and Hansen	Eytelwein	−0·83	0·36	1·19	0·81	6·7	5·0
	Hiley	−0·14	0·42	0·56	1·40	1·4	1·3
	Janbu	−·37	0·17	0·54	0·98	2·3	2·0
	Danish	−0·33	0·24	0·57	0·98	2·1	1·8
	Wave equation	−0·41	0·16	0·57	0·90	2·6	2·1
Agerschou	*Engineering News*	−0·70	0·70	1·40	0·60	5·0	4·5
	Weisbach	−0·39	0·34	0·73	0·75	2·5	2·4
	Danish	−0·34	0·23	0·57	0·90	2·2	2·0

Where the median value of μ for any formula is not equal to 1 the formula is biased. In the case of Hiley's formula the values of $\log \mu$ given in Table 4.3 at both upper and lower $2\frac{1}{2}\%$ limits are higher than for other formulae of comparable range. If the median value of μ in Hiley's formula is brought to 1·0 and the limit values of $\log \mu$ adjusted, the value of μ at the lower $2\frac{1}{2}\%$ limit becomes $-0\cdot 28$ which corresponds more closely with the other formulae.

It is often claimed by practising engineers that within a limited context of pile and soil type, some particular driving formula may rank as "good". The *Engineering News* formula for timber piles and the Hiley formula for concrete piles are examples of this. When assessing a situation and making a decision to use a particular formula the engineer is applying a personal skill which is the outcome of experience. The unselective statistical appraisal given above must be adopted in circumstances where experience is not available.

When driving a pile into a soil which dilates when disturbed, a negative pore-water pressure may be generated, with an accompanying temporary increase in the shear strength. If driving is stopped, the pore pressure is restored and the resistance, when driving is recommended, will be found to have dropped. On the other hand, driving a pile into clay generally causes an increased pore-water pressure which dissipates during a pause in driving, the soil gradually increasing in shear strength, so that on redriving the resistance is found to have increased. This is called "take up" of the pile. In most clays the permeability is too low to permit complete dissipation of pore pressures within a few hours and the resistance to driving can continue to increase for months. When redriving following a pause shows decreased resistance, however, it is prudent to make loading tests rather than place reliance on a driving formula.

Most driving formulae pay no regard to the nature of the soil, yet it is well known that no formula can be used with uniformly satisfactory results and that better agreement with test loads is obtained for piles which are end-bearing in sand or gravel than for friction piles in clay. Engineers with experience of pile driving in a particular area make modifications to formulae to obtain greater reliability, or work by some rule-of-thumb. Where piles have been driven on a site and loading tests have been made, comparison of the ultimate bearing capacity found by test with that derived from a chosen formula enables a correction coefficient to be established, which is then used with the formula for further piles driven on the site in the same soil conditions, to give some improvement in the accuracy of the calculation. A driving formula is of most practical use as a control where the ground conditions are substantially uniform to ensure that piles of one kind are driven to approximately the same resistance, within the limits of length decided upon from the site investigation results. A simple record of blow-count per foot of penetration is, of course, valuable for indicating contact with a resistant bed.

Taking the many considerations into account, the following general principles should be observed when using driving formulae:

1. If possible, avoid using a driving formula except for piles that are to be end bearing in sand or gravel.
2. Use a formula that gives a small scatter of the ratio of real ultimate bearing capacity to computed ultimate bearing capacity for the particular type of pile.
3. Use the simplest formula that meets (2) above. There is no merit in complication if it does not produce reliable results.
4. Carry out test loadings and apply a correction coefficient to the formula. The coefficient will depend upon the soil conditions, the type and size of pile, the driving equipment and its operator. A change in any one of these factors will affect the coefficient.
5. Use a factor of safety that is realistic, bearing in mind the anticipated error in the computed ultimate bearing capacity particularly if loading tests are not made to obtain a correction coefficient.

CHAPTER 5

PILE DRIVING BY VIBRATION

Introduction

From about 1935 Russian engineers have been developing methods of pile driving by vibration and a description of the Russian achievements given by D. D. Barkan to the International Society of Soil Mechanics and Foundation Engineering in London in 1957 attracted a great deal of attention, since many engineers were previously unaware of the practical possibilities of the method.

Vibratory pile drivers operate by transmitting vibrations of the requisite frequency and amplitude in the direction of the length of the pile, whereby the resistance of the soil to penetration is reduced and the pile sinks into the ground under the combined weight of the driver and the pile, or with the aid of a surcharging load or a rope pull-down.

Basically, a typical vibrator of the original Russian pattern consists of a case containing a pair of axles geared together to rotate at equal speed in opposite directions. The axles carry eccentric weights and are driven by an electric motor mounted on the case, as shown diagrammatically in Fig. 5.1. Hydraulic clamps at the bottom of the case provide an attachment to the head of the pile. In later designs the motors are enclosed inside the case and the eccentric weights are mounted directly on the motor shafts. In large vibrators more than one pair of axles carrying eccentrics are used.

In the type of vibrator employing surcharge, the vibrator unit is rigidly attached to the pile and the driving motor and additional

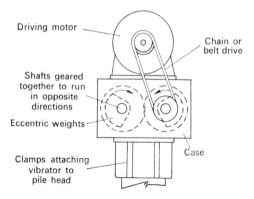

Fig. 5.1. Two-axle vibrator with the driving motor mounted on the case.

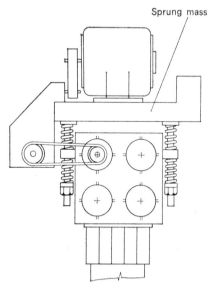

Fig. 5.2. Vibrator with the driving motor and surcharging load supported on springs.

PILE DRIVING BY VIBRATION

weights are supported on springs, so that they do not vibrate, but provide a static load, as shown in Fig. 5.2.

Russian vibrators have been made of different weights and frequencies to cover a range of duties. For driving steel sheet piles, open ended tubes and H section bearing piles, frequencies in the order of 12 to 25 c/s are used. For heavy piles with large point resistance, such as reinforced concrete piles and large diameter tubes, lower frequencies, in the order of 5 to 10 c/s are used. The amplitude of vibration to achieve effective driving ranges from 2 to 20 mm (0·079 to 0·79 in.) depending on the type of pile to be driven and the nature of the soil.

The successful use of vibration for pile driving by Russian engineers has led to the development of vibrators in other countries. Vibrators made by German manufacturers are similar in principle to the Russian pattern and operate at frequencies from about 16 to about 39 c/s. Table 5.1 gives brief details of the Muller and

TABLE 5.1

	Muller vibrators		Schenck vibrators				
	MS–26	MS–26D	DR–12	DR–60	DR–60G	DR–120	DR–120G
No. of motors	2	4	2	2	2	4	4
Rating of each motor in kW	27	27	27	40 or 50	40 or 50	40 or 50	40 or 50
Frequency c/s	24·4	24·4	24·3	17·2, 20, 24·2, 29·5, 32, 39·2	15·8, 19·2, 30·5, 37·5	17·2, 20, 24·2, 29·5, 32, 39·2	15·8, 19·2, 30·5, 37·5
Approximate weight, kg (ton)	4,700 (4·7)	7,600 (7·5)	4,270 (4·2)	7,300 (7·2)	10,160 (10)	17,300 (17)	20,320 (20)

Schenck vibrators. The frequency and amplitude of models DR-60, DR-60G, DR-120 and DR-120G may be adjusted according to the nature of the soil and the work to be done. With models DR-60G and DR-120G the vibrator units are mounted on yokes for clamping around casing tubes used for forming large diameter bored piles, so permitting access to the full cross-section of the casing for the removal of soil and for concreting.

Figure 5.3 shows a Muller MS-26D vibrator driving steel sheet piling at a site in Britain. In this model the eccentric weights are mounted direct on the motor shafts within the case.

German vibrators have been used in Britain chiefly for driving and extracting steel sheet piling and casing tubes for large diameter bored piles. They are particularly useful for the latter kind of work, since the vibration also assists the compaction of the concrete and reduces the risk of "waisting" or the formation of gaps in the shaft during extraction of the casing.

The theory of vibration driving

In Russia the theory of vibration pile driving has followed two lines of approach. Barkan, during the period 1935 to 1953, studied the effect of vibration on the internal friction in sands and found that under vibration the particles acquired mobility. He was able to reduce the problem of vibratory pile driving to that of the motion of a body in a viscous medium. On the other hand, Neimark (1953) and other workers showed that it was possible to develop a theory based on the reduction of static friction due to the effect of vibration. As an illustration of this principle, consider a body of weight X resting on a horizontal surface. The force required to cause it to slide must exceed the static friction, μX, between the body and the surface, where μ is the coefficient of friction. If the body is vibrated by an oscillatory force which is greater than μX, the frictional restraint is destroyed and a small steadily applied force will move the vibrating body across the surface. This principle, it is claimed, may be used to overcome the skin friction of a pile.

Savinov and Luskin (1960) say that Barkan's approach fails to explain certain experimental facts, while that of Neimark and other authors takes no account of the change in the properties of the soil when subjected to vibration.

In a simple two axle vibrator the horizontal components of the centrifugal force of the two rotating eccentric weights, being equal and opposed, produce no external effect. The vertical components of the centrifugal force produce the exciting force of the vibrator, which has a maximum value, F_0, given by the expression

$$F_0 = Wr\omega^2/g \tag{5.1}$$

where W = the sum of the two eccentric weights;

r = the distance of each weight from its axis of rotation;

ω = the speed of rotation in radians/sec.

If P is the total weight of the vibrator and the pile that is rigidly attached to it, then it can be shown that the maximum amplitude, a_0, when the system is vibrating without restraint is given by the expression

$$a_0 = Wr/P \tag{5.2}$$

According to Neimark (1953), the amplitude of vibration of a pile may be determined by the approximate expression

$$a = \frac{Wr}{P}\left[1 - (4R/\pi F_0)^2\right]^{\frac{1}{2}} \tag{5.3}$$

where a = the amplitude of vibration;

R = the resistance due to skin friction of the pile.

It is seen from Eq. (5.3) that when R is considerably less than F_0, so that $R/F_0 \to 0$, then a approaches the value $Wr/P = a_0$.

When the frictional resistance is overcome and the pile slips in the soil, it is found in practice that the amplitude is about the same as that of a pile vibrating without restraint.

Savinov and Luskin (1960) recommend that when driving a given pile into a given soil, using a vibrator of the Russian pattern, the following basic conditions should be satisfied:

1. The exciting force, F_0, must be large enough to overcome the skin friction on the pile, to enable the pile to reach the required depth of penetration.

2. The amplitude must be close to that at which free oscillation of the pile in the soil will occur.

3. The external force on the vibrating system must be great enough to give the necessary rate of penetration. If the total weight of the pile and vibrator (i.e. the external force) is inadequate, a surcharging load may be added, this load being supported on springs so that it is not part of the vibrating body, although its weight acts on the pile.

In vibrators which employ additional loads supported on springs, the best results are obtained when the vibrating part weighs as little and the added load as much as possible. The natural frequency of the additional load on its springs should be considerably lower than the frequency of the vibrator. In these conditions the load itself has negligible vibration.

If instead of a surcharging load a rope is attached to the sprung platform of the vibrator, a pull-down on the pile can be obtained by making use of the weight of the piling rig or crane to provide the reaction. When extracting piles, the vibrator is slung from a crane hook as shown in Fig. 5.3 and a steady upward pull is applied through springs.

Savinov and Luskin (1960) give tables of empirical values of skin friction, recommended amplitude and point pressure for calculating the parameters of vibrators for driving different types of piles into various kinds of soil, based on the principles outlined above.

Even when the skin friction on the shaft has been overcome, it is necessary with piles which have appreciable point resistance for the amplitude of the vibration to be large enough to overcome the point resistance by percussive action in dry soils, or as a result of liquefaction in saturated soils. It must be noted, however, that high point resistance is overcome only with difficulty, and because

FIG. 5.3. A B.S.P.-Muller MS–26D vibrator driving pairs of Frodingham section 4 steel sheet piling.

Fig. 5.4. The resonant pile driver installing a 355 mm (14 in.) diameter steel tube.

of this, according to Barkan (1957), the use of vibrators for driving bearing piles is of limited application, although he claims that vibration combined with jetting will be widely used in future.

Experience in Britain with German vibrators shows that they are satisfactory for work in soft clays, sands and gravels and they have been used in weathered Keuper marl. They do not operate successfully in stiff clay.

In contrast with vibrators having frequencies up to about 40 c/s, a vibrator invented by A. Bodine in America is designed to operate at the resonant frequency of the pile at its fundamental mode of vibration, or in the case of a long pile, at the frequency of some higher harmonic. The vibrator, which is termed a resonant pile driver, is shown in Fig. 5.4. It uses two 500 H.P. engines, the total weight with engines being about 10,160 kg (10 ton). The frequency range is about 60 to 130 c/s.

The resonant pile driver developed and operated by G.K.N. Foundations Ltd. in Britain will drive steel sheet piles and tubes very rapidly in granular soils and will also drive into clays. At a site in Yorkshire a 355 mm (14 in.) steel tube was driven with a detachable shoe to form cast-in-place piles through 7·6 to 10·7 m (20 to 35 ft) of firm laminated clay of shear strength 0·49 to 0·88 kgf/cm^2 (47880 to 86184 N/m^2, 1000 to 1800 lbf/ft^2) and 4·6 to 6 m (15 to 20 ft) of sandy silt of Standard Penetration Test $N = 30$ to 35 into compact sand of Standard Penetration Test $N = 50$. Driving normally took $2\frac{1}{2}$ to $3\frac{1}{2}$ min. The Standard Penetration Test will be dealt with in Chapter 6.

The natural frequency, f, of a pile of length L when vibrating longitudinally in its fundamental mode is given by

$$f = v/2L$$

where v is the speed of sound in the pile material. For steel $v = 5020$ m/s (16500 ft/s), so that in a free condition a steel pile 24·5 m (80 ft) long would have a fundamental resonant frequency of 103 c/s. When the pile is held for driving, the effective pile length is increased and the resulting resonant frequency is therefore slightly reduced.

Rockefeller (1967) has analysed resonant and non-resonant power systems with particular reference to the Bodine resonant pile driver. He considered the pile to be a series of masses joined by springs, each mass being subjected to both elastic and damping (i.e. viscous) forces, after the manner of the analysis made by Smith (1962) of a pile driven by impact. The analysis was limited by the same problems that were present in Smith's analysis, namely the lack of information about the dynamic soil parameters entering the calculations.

General comment

In practice the decision to use a vibrator, and the choice of the particular vibrator for a given job has in most cases been largely a matter for trial and error, since the soil parameters cannot usually be defined with the accuracy necessary for assessment by calculation. Often a vibrator may be brought to a site to overcome a specific problem, as for example where casing tubes or sheet piling are difficult to extract by normal apparatus. This experience often leads the engineer to experiment with further uses of vibration.

Vibratory impact drivers

If the vibrator were not rigidly attached to the pile head it would bounce up and down giving blows to the pile at a frequency dependent on many factors. By connecting the vibrator to the pile by springs as shown diagramatically in Fig. 5.5, the blows may be controlled so that they are delivered at the same, one half, or some other fraction of the frequency of the vibrator.

Using this principle, Russian engineers have developed vibratory impact drivers which have been described by Smorodinov *et al.* (1967). Modern piling plant in Russia is commonly driven by electricity and it seems that the development of vibratory impact machines has been influenced by the availability of electric power on Russian construction sites.

Beneath the case of the vibrator is a spherically faced hammer which impinges on an anvil on the pile. The springs supporting the vibrator may be adjusted so that when at rest the hammer is

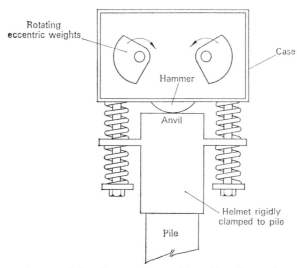

FIG. 5.5. Diagram of the vibratory impact driver. The vibrator bounces up and down, giving blows to the pile.

clear of the anvil, just in contact, or pulled down onto the anvil, and the performance of the driver depends on this adjustment and on the stiffness of the springs.

According to Smorodinov *et al.* (1967) Russian vibratory impact drivers ("Vibro hammers") vary in weight from 150 to 2300 kgf (330 lbf to $2\frac{1}{4}$ tonf); frequencies of impact are from 480 to 1450 blows/min and the power of the motors from 2 to 28 kW. With a pile having large point resistance an impact frequency of 400 to 750 blows/min is most suitable. Because the vibrator has, in fact, become a hammer, its effectiveness depends on the ratio of its weight to that of the pile. For lightweight piles the ratio of vibrator weight to pile weight is normally 1:1, for reinforced concrete piles, it is not less than 1:2. In its present state of development the vibratory impact driver is chiefly of use for rods, probes, timber piles and small concrete piles.

CHAPTER 6

THE CALCULATION OF THE ULTIMATE BEARING CAPACITY OF A PILE FROM SOIL PROPERTIES

Introduction

The calculation of the ultimate bearing capacity of a pile from measured soil properties is a logical step in the application of the principles of soil mechanics to piled foundations and the efforts of engineers in this direction have resulted in a number of "static" formulae, so called by comparison with the "dynamic" formulae considered in Chapter 4.

The following factors are responsible for the load-carrying ability of a pile:

1. As the base, or point, of a pile is pressed downward by a load on the head, the soil immediately below and to the sides of the base must be pushed aside. The soil offers a resistance to the shearing action which such a movement entails.

2. Downward movement of the pile relative to the soil surrounding it causes the mobilisation of tangential forces on the shaft surface that oppose motion. These forces are due both to adhesion and to the friction of the soil on the shaft surface.

3. The pile takes the place of a certain volume of soil, the weight of which was previously carried by the soil below the base of the pile. If the depth of the base is D and the average density of the soil in this depth is $\bar{\gamma}$, then each unit area in the horizontal plane at the level of the pile base was supporting a pressure of D in equilibrium before the pile was introduced. If the pressure on the

CALCULATION OF THE ULTIMATE BEARING CAPACITY

soil beneath the base of the pile is greater than $\bar{\gamma}D$ when the pile is in place, then the additional pressure on the soil beneath the base must be taken in the manner described in 1.

Taking these three factors into account, the ultimate bearing capacity of the pile is generally assumed to be reached when

$$P_u + W = f_u A_s + q_u A_b + \bar{\gamma} D A_b \tag{6.1}$$

where P_u = the applied load at ultimate bearing capacity;
W = the weight of the pile;
f_u = the ultimate value of the tangential force per unit area of the shaft due to adhesion and skin friction;
q_u = the ultimate value of the resistance per unit area of the base due to the shearing strength of the soil;
A_s = the area of the shaft;
A_b = the area of the base of the pile.

(Observe the changed notation which will be used in this and further chapters.)

Very often both W and the term $\bar{\gamma} D A_b$ are neglected, or they are assumed to be equal. Equation (6.1) then becomes

$$P_u = f_u A_s + q_u A_b \tag{6.2}$$

The values given to q_u and f_u may be found experimentally, or calculated from measured soil properties, or may have their origin in rule of thumb.

The approach outlined above forms the starting point of almost all static pile formulae, since with hardly any exception it is assumed that the components of resistance may be separately evaluated and that the sum of their ultimate values gives the ultimate bearing capacity of the pile.

The analytical problem presented by a pile installed in real soil and then subjected to loading is one of extreme complexity. In the attempts to find solutions suitable for practical design purposes, simplified pile-soil systems having limited and arbitrary properties have been posed, and the resulting "models" often bear only slight resemblance to reality. For example, it is implicit in most formulae that the pile is created in position without affecting the soil in which it is placed. Very little is known experimentally about the distribution and magnitude of the soil and pore water pressures on the shaft and on the base, or in the soil when a pile is first installed and during the period prior to loading when the changes causing "take up" occur. Similarly, the changes which take place as the pile is progressively loaded until it penetrates further into the ground are almost unknown.

This lack of knowledge of many of the phenomena which form the very essence of pile behaviour precludes the creation of a formula based on reality. With the advance of knowledge in a general way about soil properties, engineers are better able to appreciate the dangers of undue simplification and the limited applicability and approximate nature of formulae. Clearly, the more factors known to influence pile behaviour that can be evaluated and brought with in the compass of a "formula" the greater the potentiality of that formula for dealing with a variety of circumstances.

Early formulae

In the second half of the nineteenth century when engineers were attempting to solve problems by the use of applied mechanics, the "soil" considered was idealised as a uniform, dry, granular material. The facts that the granular soils like sand could compact or dilate and were often saturated, were not taken into account, while the quite different properties possessed by clay were generally ignored.

Rankine's theory of conjugate stresses for a material having internal friction formed the starting point of some bearing capacity theories. According to Rankine, if σ_1 and σ_3 are the conjugate major and minor principle stresses when the state of plastic equilibrium is reached in a soil with an angle of internal friction ϕ then

$$\frac{\sigma_1}{\sigma_3} = \frac{1 + \sin\phi}{1 - \sin\phi} \qquad (6.3)$$

CALCULATION OF THE ULTIMATE BEARING CAPACITY

According to Paton (1895), who applied Rankin's theory of conjugate stresses to obtain a solution for the bearing capacity of a pile, the horizontal pressure on the pile shaft at a depth D in soil of density γ will be between the values

$$\gamma D \frac{(1+\sin\phi)}{(1-\sin\phi)} \quad \text{and} \quad \gamma D \frac{(1-\sin\phi)}{(1+\sin\phi)},$$

depending on whether γD is the minor or the major principle stress. If the lower value is taken for safety, the average pressure on a shaft of length L will be

$$\frac{\gamma L(1-\sin\phi)}{2(1+\sin\phi)}$$

and the total pressure will be

$$\frac{\gamma L(1-\sin\phi)}{2(1+\sin\phi)} A_s,$$

where A_s is the surface area of the shaft. If the coefficient of friction of the soil on the shaft is μ, the total skin friction which could be developed at the shaft surface is

$$\frac{\mu\gamma L(1-\sin\phi)}{2(1+\sin\phi)} A_s \tag{6.4}$$

To obtain the ultimate bearing capacity of the base, the equilibrium conditions of the two small elements of soil B and C in Fig. 6.1 are determined. When the pile base reaches its ultimate bearing capacity, the element B has a vertical pressure q_u. If the horizontal pressure between B and C is p, then

$$\frac{q_u}{p} = \frac{(1+\sin\phi)}{(1-\sin\phi)}$$

q_u being assumed to be the major principal stress.

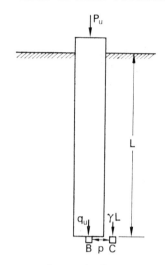

Fig. 6.1. The stresses on elemental cubes of soil at the base of a pile.

At C, if the element is in equilibrium under a major principal stress p and the vertical stress γL, then

$$\frac{p}{\gamma L} = \frac{(1 + \sin\phi)}{(1 - \sin\phi)}$$

Thus

$$q_u = \gamma L \frac{(1 + \sin\phi)^2}{(1 - \sin\phi)^2} \qquad (6.5)$$

The total load P_u which may be supported by the pile is thus

$$P_u = \mu\gamma \frac{L(1 - \sin\phi)}{2(1 + \sin\phi)} A_s + \gamma L \frac{(1 + \sin\phi)^2}{(1 - \sin\phi)^2} A_b \qquad (6.6)$$

where A_b is the area of the base.

Other engineers also developed expressions for the bearing capacity of piles in terms of ϕ, but none gave results that could be relied on when applied to practical cases, and the early "static" formulae fell into disrepute.

CALCULATION OF THE ULTIMATE BEARING CAPACITY

The principal source of error was in the value given to ϕ. At this period most engineers thought of each soil, whether sand or clay, as having an "angle of repose" that was considered to be the same as Rankine's ϕ. Soils would be identified by observation and an appropriate value of ϕ allocated from an accepted set of values then current. Very little was understood about the real nature of "internal friction" or of its measurement for use in calculations.

Recent formulae

Various theoretical solutions were proposed for the two dimensional problem of the bearing capacity of shallow strip footing from about 1934 onward, Terzaghi's solution published in 1943 being well known. Terzaghi assumed that failure was the result of the shearing and upward heaving of the soil at each side of the foundation, as in Fig. 6.2, and he extended his solution to square

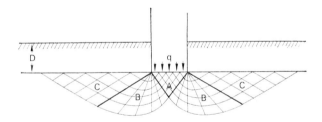

FIG. 6.2. The zones of shear beneath a shallow foundation according to Terzaghi. A, zone of elastic equilibrium; B, zones of radial shear; C, zones of passive shear.

and circular shallow foundations by introducing coefficients derived empirically.

For a circular foundation near the surface, Terzaghi gives

$$q_u = 1.3cN_c + \gamma D N_q + 0.6 \frac{B}{2} \gamma N_\gamma \qquad (6.7)$$

where q_u = the ultimate bearing capacity per unit area
  c = the cohesion of the soil
  D = the depth of the foundation
  B = the breadth of the foundation
  γ = the density of the soil
N_c, N_q and N_γ = bearing capacity factors which are dependent only on ϕ, the angle of internal friction of the soil.

Terzaghi discussed the mechanism of pile behaviour and indicated the factors contributing to the resistance of the base. Thus in Fig. 6.3 the downward movement of the base AB causes soil

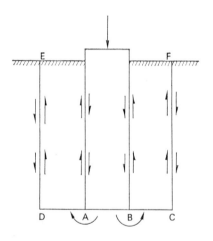

FIG. 6.3. Diagram showing the shear forces resisting heaving of the soil at the base of a pier or pile, according to Terzaghi.

to be displaced outward and upward. This is resisted not only by the weight of an annulus of soil represented in cross-section by *CDEF*, but also by shear forces opposing upward movement of the annulus along its outer surface *CF* and *DE* and on the surface of the pile shaft.

CALCULATION OF THE ULTIMATE BEARING CAPACITY

Terzaghi did not pursue a solution for this model, because of uncertainty about the distribution and values of the shear forces in the soil and at the shaft surface. For practical design purposes Terzaghi and Peck (1967) suggest that for piles driven through compressible material to a firm base, the base resistance should be calculated from Eq. (6.7). The shaft resistance should be taken as $f_u A_s$, where f_u, the average ultimate skin friction per unit area, should be determined by experiment.

Solutions have been put forward for the best resistance of a pile using Terzaghi's suggestion that a surcharge effect exists due to the frictional resistance on the shaft, but none have come into practical use for lack of adequate knowledge of the stress field around a pile and the consequent reliance on assumptions.

Berezantzev, Krisoforov and Golubkov (1961) postulated a mechanism for a driven pile in a granular soil in which the zones of shear, when failure at the base is reached, do not progress above the level of the base, and found solutions for the stress field by the theory of limit equilibrium in a granular medium (Berezantzev 1952). The surcharge p_0 at the level of the base is taken as the weight of an annulus of soil of certain radius surrounding the pile, less the friction on the outer vertical surface of this annulus. The average value of $p_0 = \alpha_T \gamma_D D$, where α_T is a coefficient depending on D/B as well as on ϕ and γ_D is the value of γ at depth D. The ultimate bearing capacity per unit area of the base is given by

$$q_u = A_k \gamma B + B_k \alpha_T \gamma_D D \tag{6.8}$$

The radius of the annulus of soil and the lateral soil pressure on its outer surface and the values of the coefficients in Eq. (6.8) are calculated from the theory of limit equilibrium in a granular medium.

The failure mechanism for the base of a deep foundation used by Meyerhof (1951) in his general theory of bearing capacity, provides a working basis for explaining some of the known phenomena in pile behaviour. Meyerhof assumes a system of shear zones as in Fig. 6.4 for the case of a deep foundation with rough surfaces. Below the base is a central zone *ABC* which remains in an elastic state of equilibrium and acts as part of the foundation; on each side of this zone there are two plastic zones *ACD* and *BCE* (i.e. zones of radial shear) and two zones of plane or mixed shear, *DAF* and *EBG*.

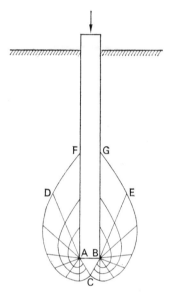

FIG. 6.4. The zones of shear around the base of a pile, according to Meyerhof.

Meyerhof expressed the base resistance per unit area as

$$q_u = cN_c + K_s \gamma D N_q + \frac{B}{2} N_\gamma \qquad (6.9)$$

where K_s = the coefficient of earth pressure on the shaft within the failure zone; varying from $\frac{1}{2}$ for loose soil to about 1 for dense soil;

N_c, N_q and N_γ = bearing capacity factors that are dependent on ϕ and the embedment ratio $\dfrac{D}{B}$.

CALCULATION OF THE ULTIMATE BEARING CAPACITY

In a soil giving both adhesion and friction on the shaft of the pile, Meyerhof (1953) expressed the tangential force per unit area as

$$f_u = c_a + K_s \gamma D \tan \delta$$

where c_a = the adhesion per unit area;
δ = the angle of friction of the soil on the shaft.
For clay, $\delta = 0$, so that $f_u = c_a$.
For non-cohesive soil, $c_a = 0$, so that $f_u = K_s \gamma D \tan \delta$.

The actual values of f_u, c_a, K_s, γ and δ probably vary from point to point down the shaft, but for practical purposes average values are adequate.

Thus, Meyerhof's formula for the bearing capacity of a pile in a soil possessing both cohesion and friction is

$$P_u = A_s(c_a + K_s \gamma D \tan \delta)$$
$$+ A_b\left(cN_c + K_s \gamma D N_q + \frac{B}{2} N_\gamma\right) \quad (6.10)$$

In the case of a pile of normal proportions D/B is about 30 or more, so that in the above solutions for q_u (Eqs. (6.7) (6.8) and (6.9)), the term involving B will be small in relation to the other terms and may be ignored. The equations then become

Terzaghi $\qquad q_u = 1\cdot 3\, cN_c + \gamma D N_q \qquad (6.11)$

Berezantzev *et al.* $\quad q_u = B_k \alpha_T \gamma_D D \qquad (6.12)$

Meyerhof $\qquad q_u = cN_c + K_s \gamma D N_q \qquad (6.13)$

Equations (6.11) and (6.13) apply to soil having both cohesion and internal friction, Eq. (6.12) is applicable only to a frictional soil.

Practical methods of calculating the ultimate bearing capacity

For practical purposes when calculating the shaft and base resistances of a pile, soils are regarded as either entirely cohesive or entirely non-cohesive. It is convenient to deal separately with the behaviour of piles in each type of soil and piles that are supported by rock.

PILES THAT ARE END-BEARING IN SAND OR GRAVEL

When a pile penetrates weak strata to obtain end-bearing in a bed of non-cohesive soil such as sand or gravel, the major part of the ultimate bearing capacity is generally supplied by the base resistance. Shaft friction may contribute to the bearing capacity, but in the case of a weak soil that is in the process of consolidation, the soil may in the long term cause a downward drag on the shaft. This latter condition will be examined in Chapter 8.

For a pile with its base in non-cohesive soil, the terms containing c in Terzaghi's and Meyerhof's equations for q_u become zero, since $c = 0$. Equations (6.11), (6.12) and (6.13) then become

Terzaghi $\qquad q_u/\gamma D = N_q \qquad (6.14)$

Berezantzev et al. $\qquad q_u/\gamma_D D = B_k \alpha_T \qquad (6.15)$

Meyerhof $\qquad q_u/\gamma D = K_s N_q \qquad (6.16)$

Clearly, the quantities N_q, $B_k \alpha_T$ and $K_s N_q$ may be compared, as in Fig. 6.5, in which the curves represent values of N_q given by Terzaghi (1943), values of $B_k \alpha_T$ from data given by Berezantzev et al. (1961) and values of $K_s N_q$ calculated from data given by Meyerhof (1953) assuming $K_s = \frac{1}{2}$. The values of α_T for $D/B = 50$ used in the calculation have been obtained by extrapolation of the data.

As the base of a displacement pile of diameter B is forced into sand, the progression of the failure zone causes shearing within a cylinder of diameter a, and compaction of the sand within a cylinder of larger diameter b, as in Fig. 6.6. For loose sand, Meyerhof (1959) found $a = 4B$ and $b = 6B$ to $8B$. For dense sand, Kerisel (1961) found $a = 3B$ and $b = 5B$.

Most theories do not take into account the compaction of a sand resulting from pile driving, and the value of N_q or other comparable coefficient of γD in the equation for q_u, is that applicable to the ϕ of the soil in its unaffected state prior to piling.

If agreement is found between such a theory and the results of loading tests, one or other of two things must occur. Either the

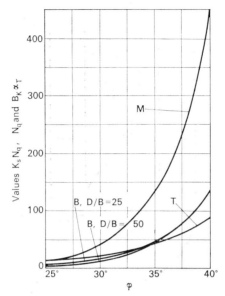

Fig. 6.5. Curve T gives N_q versus ϕ according to Terzaghi. Curves B give $B_k \, a_T$ versus ϕ according to Berezantsev *et al.* for $D/B = 25$ and 50. Curve M gives $K_s N_q$ versus ϕ, according to Meyerhof for $K_s = 1/2$.

effect of compaction is negligible, or the theory by which N_q or other coefficient is calculated is in error by an amount which compensates for the use of a value of ϕ that is lower than that of the compacted soil actually surrounding the pile. Meyerhof (1959) has attempted to take account of the compaction resulting from the installation of displacement piles.

The weakness of the current theories, as of the early static theories, is that values of ϕ, δ, K_s or other soil parameters are required and the value of a theory for practical application is governed by the accuracy with which the necessary parameters can be determined.

The difficulty and expense of extracting sand in an "undisturbed" state from boreholes in order to determine the angle of internal friction and other properties, the doubtful accuracy of the

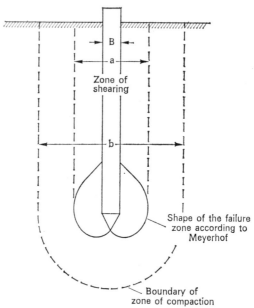

Fig. 6.6. The zones of shearing and compaction around a driven pile.

results and the impossibility of dealing with gravels in this way, have led to the development of penetrometers.

"Dynamic" penetrometers are driven into the soil by blows of a hammer, and "static" penetrometers are forced down by steady pressure, to give information about the density or other properties of the soil encountered. The best known and most widely used of the dynamic methods is the so-called "standard penetration test" and of the static methods, that commonly called the "Dutch cone" penetrometer test.

The standard penetration test

The test consists of driving a split-barrel sampler of 2 in. (50·80 mm) outside diameter into the soil at the bottom of a cased borehole. The sampler is attached to drill rods and driving is by blows from a 140 lbf (63·5 kgf) weight falling 30 in. (762 mm) onto the top of the rods. The sampler is driven a total distance of 18 in. (457 mm) and the number of blows needed to drive the final 12 in. (305 mm) of this distance is termed the standard penetration value N. Great care is required in making the borehole for the test to avoid disturbance of the soil at the bottom of the hole, since any loosening of the soil reduces the N value. The test is covered by British Standard 1377–1967.

Terzaghi and Peck (1967) have correlated N values with descriptions of the density of sand deposits and with soil pressures causing 1 in. (25·4 mm) settlement of footings of different widths. Peck, Hanson and Thorburn (1953) have given a correlation of N with the angle of internal friction ϕ and Terzaghi's bearing capacity factors N_q and N_γ. This is reproduced in Fig. 6.7.

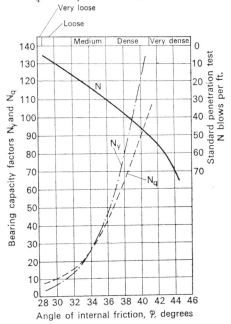

FIG. 6.7. Correlation of the standard penetration test N values with N_q, N_γ and ϕ.

When the test is made in fine sand or silt below the water table a correction is generally made by which the blow-count is reduced. If N' is the S.P.T. value as found, then the equivalent value of N to be taken for design purposes when N' is greater than 15 is given by the expression

$$N = 15 + \tfrac{1}{2}(N' - 15)$$

Gibbs and Holtz (1957) have introduced a further correction to the N value obtained at shallow depths to allow for the effect of the overburden pressure at the level of the test. With an overburden of about 15 m (50 ft) no correction is required, so that for piling the Gibbs and Holtz correction may generally be ignored.

Many practising engineers claim to use the standard penetration test for predetermining the end-bearing capacity of driven piles. Both the test and the driving of piles rely on impact and in each case the driving resistance is related to the density of the soil and the overburden pressures, but there is practically no evidence of direct correlation to be found in the literature. Until a direct correlation becomes available, a method of estimating the ultimate resistance of the base of an end-bearing pile in sand from standard penetration test results is by means of the correlation of N with N_q given in Fig. 6.7. The value of q_u is then determined from $q_u = \gamma D N_q$.

In dry sand the values of N found at different test points in the bed will vary due to random variations in the density of the deposit and the lowest value at or within a few feet below the level of the pile base should be taken in design.

When in sand below the water table, the borehole must be maintained full of water both during boring and testing to minimise disturbance due to inflow at the bottom and the N values obtained may be more representative of the practical difficulties and the operational technique employed in making the borehole than of the original density of the sand deposit. Whether in any given case a standard penetration test result can be relied upon must depend on the engineer's assessment of the circumstances on the spot.

CALCULATION OF THE ULTIMATE BEARING CAPACITY

In order to extend the applicability of this type of test to sandy gravels and gravels, Palmer and Stuart (1957) fitted a solid conical tip to the sampler and found that the results were comparable with those obtained with the standard apparatus.

A method of estimating the approximate working load on timber friction piles is used by the American Bureau of Reclamation which relies on a correlation between the penetration resistance of the sampler and the driving resistance of the pile (see Holtz, 1961).

The Dutch cone penetrometer

The Dutch deep sounding apparatus was originally devised for locating resistant strata to which piles might be driven. In its simplest form it consists of a probe with a conical point of base area 10 cm^2 (35·7 mm diameter) at the end of a rod sliding in a tube, as in Fig. 6.8. The tube has an external diameter equal to that of the base of the cone. In use, the assembly is pushed into the ground and the cone advanced a short distance, as in Fig. 6.8(b);

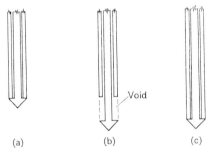

Fig. 6.8. The simple form of Dutch penetrometer. (a) As pushed into the soil ready for use. (b) 1st stage; the cone advanced. (c) 2nd stage; the tube advanced to join the cone.

the force to do this being measured. The tube is then advanced to join the cone, as in Fig. 6.8(c); the force to do this also being measured. Modifications have been made to prevent soil entering the lower gland and to permit simultaneous advancement of the cone and tube with the separate measurement of the two resistances. It is common Dutch practice to take a reading of cone and tube resistances at intervals of 20 cm in depth and to plot the results of the test as in Fig. 6.9.

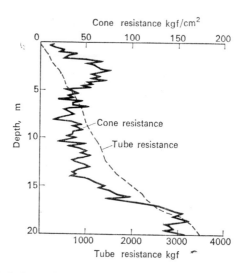

FIG. 6.9. Typical Dutch penetrometer test result, showing cone and tube resistances plotted every 20 cm of depth.

The methods by which the bearing resistance of a pile may be calculated from cone penetrometer tests have been modified with increasing experience. In the direct method, the ultimate point resistance of a pile is calculated by multiplying the cross section of the base of the pile by the measured cone resistance per unit area at the same level. The unit friction along the tube would normally be lower than the unit skin friction acting on a pile, because the larger pile causes greater compaction than a penetrometer, so that by multiplying the tube resistance by the ratio of the circumference of the pile to that of the tube, a value for the pile shaft friction is obtained which is generally on the safe side.

CALCULATION OF THE ULTIMATE BEARING CAPACITY   73

Van der Veen (1957) allowed for the fact that the shear surfaces, formed around a pile base as the ultimate bearing capacity is reached, extend for distances ad above and bd below the level of the base, where d is the equivalent diameter of the pile base (cf. Meyerhof's theory). It is assumed that the resistance of the base per unit area is the average of the measured cone resistance over the range $(a + b)d$. Van der Veen found the most probable values were $a = 3\cdot 75$ and $b = 1$. The principle is indicated in Fig. 6.10.

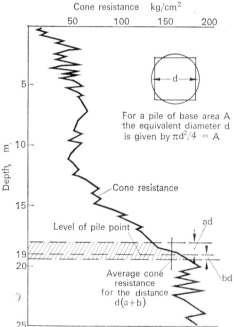

FIG. 6.10. Determination of the average cone resistance of the soil involved in shearing failure at the base of a pile of equivalent diameter d. The shear zones extend for distances ad above and bd below the base.

One serious difficulty in these simple methods of calculating the ultimate bearing capacity of the base of a pile from Dutch cone penetrometer tests is that they do not recognise any effect due to the difference in diameter between the penetrometer and the pile, i.e. the "scale" effect. De Beer (1963) has dealt with this problem and has presented a method of correcting for the scale effect which has been used for several years in Holland and Belgium. If the pattern of the shear zones given by Meyerhof in his general theory of bearing capacity for deep foundations is accepted, then for a given distance of penetration into the resistant bed the shear zones associated with a small probe of 35·7 mm diameter may possibly be entirely within the bed as in Fig. 6.11(a). For the same distance of

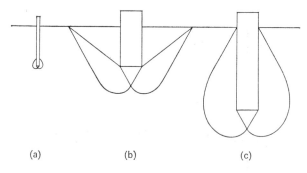

Fig. 6.11. (a) Shear zones around the cone of a penetrometer. (b) Shear zones at the base of a pile at the same depth may reach the surface of the resistant bed. (c) A greater penetration is required to ensure that the shear zones at the base of the pile do not reach the surface of the resistant bed.

penetration the shear zones of a pile (which may be ten times the diameter of the probe) will be as shown in Fig. 6.11(b). To obtain a failure system similar to that of the probe the pile would have to penetrate to the depth shown in Fig. 6.11(c). Thus, if the same bearing capacity law is to apply to both the pile and the probe, their penetrations into a resistant bed must be governed by their scale ratio.

However, the relationship is not one of scale alone, because the greater depth to which the pile has to penetrate to ensure that the shear zones do not reach the surface of the bed results in an increased overburden pressure.

De Beer (1963) has given a practical method of using the cone penetration diagram for pile design, allowing for the scale effect. When the sounding is made with a reading every 20 cm, the entry of the cone into the surface of a resistant bed is shown by three, four or more points in a line, as at AB in Fig. 6.12, as more and

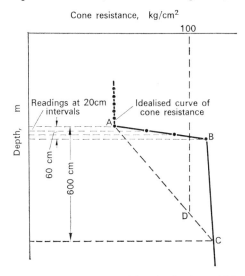

FIG. 6.12. De Beer's method of determining the depth to which a pile should penetrate to correct for the scale effect.

more of the shear zone system becomes enclosed in the bed, followed by a succession of points showing substantially the same resistance as at B. Point B represents the condition of full enclosure of the shear zones in the resistant bed. If four points are in line from A to B, then the embedment of the cone at B is 60 cm. Thus, for a pile 10 times the diameter of the cone, the depth of embedment to ensure full enclosure of the shear zones would be about 600 cm and the increase of resistance shown by the pile base would approximately follow a line such as AC, where C has a somewhat greater resistance than B due to the increased overburden.

Suppose it is decided that the pile is to be driven until the base resistance is 100 kgf/cm^2, since to attempt to reach a higher value might cause damage to the pile. This point is represented by D on AC and the level of D gives the depth to which the pile point should be driven. A pile capable of withstanding the forces involved could be driven to the level of point C, to obtain the maximum resistance, provided the under surface of the bed was sufficiently far below the pile base to ensure that the shear zones were still within the bed.

In the original design of the Dutch penetrometer, the force to advance the tube represents a summation of the effects due to soils of different types between the ground surface and the cone. In order to measure the frictional resistance of the successive beds

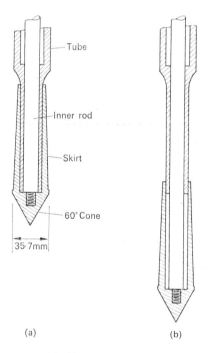

FIG. 6.13. Penetrometer with skirted cone; (a) as installed in the soil, (b) with the cone advanced.

CALCULATION OF THE ULTIMATE BEARING CAPACITY 77

of soil separately, Begemann (1953) used a sleeve about 13 cm long attached immediately above the cone in such a manner that the force on the sleeve could be determined.

One of the difficulties with the simple penetrometer shown in Fig. 6.8 is that any soil that falls into the space left when the cone is pushed forward alone may enter the gland between the rod and the tube when the tube is advanced to join the cone. This is overcome by fitting the cone with a skirt, as in Fig. 6.13.

The reliability of the Dutch cone penetrometer for predicting the base or point loads on piles has been examined by Huizinga (1951) and Van der Veen (1953, 1957). Huizinga determined the point resistance at ultimate bearing capacity for 29 piles by subtracting the value found for the pulling resistance from the ultimate bearing capacity. With two exceptions the point resistances per unit area were between 70 and 150 per cent of the cone resistance values. He also compared the frictional values per unit area found in pulling tests with those determined from the tube of the penetrometer and found that the pile friction values were on average 200 per cent of the tube values, the variation covering the range 100 to 400 per cent.

Van der Veen (1953) gave the results of 19 comparisons between ultimate pile point resistances and cone resistances and later supplemented these (1957) with 14 more. The scatter of results coincided closely with those of Huizinga.

Other types of penetrometers

There are various other forms of both dynamic and static penetrometers in use and one or other may be favoured in a particular country. As a result of the many types and dimensions of penetrometers and the attempts to find correlations between them and with soil parameters, the present state of the art is one of confusion.

The penetrometer tests most likely to be used in British and American site investigation work are the standard penetration test and the Dutch cone penetrometer test. Should another form of test be suggested as a method of site investigation, the engineer would be well advised to examine its authority and the available correlation data before using the results for design purposes.

PILES IN CLAY

When clay is stressed the pore water pressure increases and if the stress is increased sufficiently rapidly so that failure takes place with very little dissipation of the pore pressures, a "total stress" or $\phi = 0$ type of failure analysis may be made. The strength parameter involved in these circumstances is the undrained cohesive strength, which may be determined either by the unconfined compression test or the undrained triaxial test. The sampling of clay in the field in a relatively undisturbed condition and the subsequent testing of specimens in the laboratory, present few difficulties when compared with sands or gravels and it is usually possible to prepare a plot or "profile" of cohesion versus depth for a clay stratum.

Although the long-term behaviour of piles indicates that pore pressure dissipation and consolidation in the clay around a pile do occur, no generally accepted design method based on an effective stress approach is available for practical use, although Chandler (1966, 1968) has put forward an approach to such a method. Very little experimental work has been done to find the pore pressures present in the soil around a pile, or the way the pressures dissipate.

Load carried by the base of a pile in clay

In the case of a pile of normal proportions embedded entirely in clay, the contribution to the ultimate bearing capacity given by the base or point may not be greater than 10 per cent and is often neglected in practical design. With large diameter bored piles, however, particularly those with enlarged bases, the base resistance may exceed the resistance due to shaft friction and calculation of the base resistance becomes important.

When $\phi = 0$, then $N_q = 1$, $N_\gamma = 0$ and $K_s = 1$ (Meyerhof, 1951), and assuming that the coefficient 1·3 in Eq. (6.7) may be

CALCULATION OF THE ULTIMATE BEARING CAPACITY   79

taken into the value of N_c, then both Terzaghi's and Meyerhof's equations (Eqs. (6.7) and (6.9)) for q_u at the base of a pile in cohesive soil may be reduced to the form

$$q_u = c_b N_c + \gamma D \qquad (6.17)$$

where c_b is the cohesion at the level of the base.

The value of N_c in Eq. (6.17) has been determined by analytical and experimental methods. Meyerhof obtained $N_c = 9.3$ or 9.8, analytically, the value depending on whether the base was smooth or rough. Wilson (1950), by a different approach, obtained $N_c = 8.0$, while Gibson (1950) found a value $N_c = 8.5$ by assuming that the penetration of the base at failure was equivalent to expanding a spherical hole the same diameter as the base in the clay. Skempton (1951), taking account of the various theoretical values and others found by experiments on models, concluded that the semi-empirical value $N_c = 9$ was probably sufficiently accurate for practical purposes and this value is generally accepted in design in Britain.

Driven piles

When a pile is driven into clay, shear surfaces associated with the base are progressively formed deeper and deeper in new soil as penetration proceeds. Around the shaft the soil is compressed and moved laterally and vertically to accommodate the pile, the ground surface heaves and the clay in the immediate vicinity of the pile shaft is completely remoulded. According to Casegrande (1932), the zone of remoulding has a diameter twice the diameter of the pile and the soil is sufficiently affected within a zone four times the pile diameter to cause an increase in compressibility. If the clay is sensitive, there is an immediate loss in strength due to remoulding, and in both sensitive and non-sensitive unfissured clay there is an increase in pore water pressure in the compressed zone. In the period following driving, pore pressure dissipation and drainage may be sufficient to restore the strength. This phenomenon is called "take up" and in some soils the bearing capacity of a pile relying chiefly on shaft friction may increase to many times its value immediately following driving.

This property has been demonstrated by making repeated loading tests at intervals on the same pile, obtaining a curve of ultimate bearing capacity versus time. Such a curve will only apply to the particular site and type of pile on which the test is made, for although various attempts have been made to obtain correlation with tests on soil specimens, no general theory has emerged that enables a designer to predict the bearing capacity of a pile some time after driving, with the necessary assurance, from soil tests alone.

In most cases the increase of bearing capacity with time is rapid at first, so that by the end of a month the ultimate bearing capacity is not much less than that which would be reached in a year or more. Peck (1958) made a collection and analysis of test results on behalf of the American Highway Research Board and concluded that in soft and medium clays, having unconfined compressive strengths up to about 0·98 kgf/cm^2 (2000 lbf/ft^2, 95760 N/m^2) the contribution made by shaft friction to the ultimate bearing capacity, after a period for "take up" was equal to the product of the embedded area and the original shearing strength of the soil. The shearing strength was taken to be half the unconfined compressive strength of undisturbed samples. Thus, for soft clays $f_u = c$ for design purposes.

In the case of piles driven into stiff clays, that is clays having unconfined compressive strengths of over 0·98 kgf/cm^2 Peck (1958) found that the shaft resistance was smaller than the product of the embedded area and the shearing strength of the soil and that as the shear strength increased the difference became greater. Others have also noted this effect. Thus, for stiff clays $f_u < c$.

In the stiff fissured clays, it is possible that the displacement of the clay may break it into blocks or fragments and the effect on pore pressure is not known.

A number of experiments have been made with the object of finding the load in a pile at chosen sections down its length. By attaching strain gauges at the required positions to steel piles, the strain under load may be measured and the stresses in the pile at those points may be approximately found, either directly, or from a previous callibration under load. The difference in the loads at two cross-sections represents the load carried by friction or adhesion on the surface of the shaft between the two sections (Seed and Reese, 1957). Most experimental work of this kind has been concerned with the adhesion of soft clays and the load distribution in the shaft of a model pile driven in clay given in Fig. 6.14 shows a pattern which is typical of that found in other tests on larger piles.

The effect of the material of the pile on skin friction has been examined by various people. The data have generally been in the form of collections of results of tests on concrete, steel and timber piles drawn from many sources, the piles being in soils of different

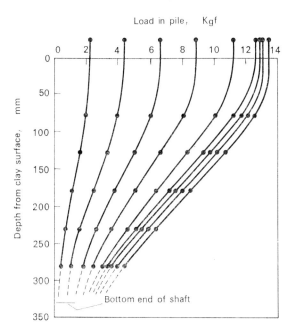

FIG. 6.14. The distribution of load down the shaft of a model pile in clay for different applied loads.

cohesive strengths. Although some authors suggest the skin friction on some one material to be greater than on others, the scatter of the results would seem to be too great to permit such a differentiation between the various pile materials.

Bored piles

For convenience, bored piles are usually separated into "normal" bored piles, that is piles up to about 610 mm (2 ft) diameter and "large" bored piles which are of 610 mm (2 ft) diameter and over and this convention will be observed.

Bored piles are most advantageously formed entirely in a reasonably uniform deposit of firm or stiff clay similar to a boulder clay or London Clay. The bore hole would be put down by an auger or by percussion with a core cutter. At present there are no recorded data to show if there is any difference between the piles resulting from the two methods of hole construction, and the choice is dictated by economic factors.

The relief of stresses around the bore hole causes migration of pore water from the surrounding clay to the surface of the bore hole, and the disturbance of fissures and the intersection of silty inclusions may also release some water. Deterioration of the sides of the bore hole from these causes can be minimised by boring and concreting as rapidly as possible. It is not uncommon for water to be poured into the hole to make boring easier and this should be discouraged. The clay has also the opportunity to take up water from the wet concrete placed in the hole to form the pile, but there is not much evidence to show how much water is taken up. Meyerhof and Murdock (1953) found there was an increase of 4 per cent in the water content of the clay in contact with the concrete in a bored pile which took 3 days to complete; the clay 76 mm (3 in.) away from the concrete being unaltered in water content.

Whatever the cause, this increase in water content reduces the strength of the clay and its adhesion on the pile shaft when the concrete has set is lower than the original value of cohesion. The com-

CALCULATION OF THE ULTIMATE BEARING CAPACITY   83

plex system of effective stresses and pore pressures within the clay and at the clay/concrete interface has not been studied, and the only design approach which is possible at the present time is by the empirical determination of f_u.

If with very soft clays the adhesion is accepted as being equal to the cohesion, then the lower limit to the value of f_u following absorption of water will be equal to the cohesive strength of the clay in the "fully softened" condition.

Ward and Green (1952) suggested a design equation for short bored piles up to about 4·25 m (14 ft) long, such as might be used for house foundations, assuming the worst conditions likely to occur:

$$P_w = 9c_b A_b/2 + \pi d c_s L_e \qquad (6.18)$$

where P_w = the working load;
c_s = the fully softened shear strength;
d = the diameter of the shaft;
L_e = the effective length over which adhesion occurs.

An allowance is made in calculating the value of L_e for the clay shrinking away from the upper end of the pile in dry weather, no adhesion being assumed on the upper 1·22 m (4 ft) where the vegetation is grass, or on the upper 2·13 m (7 ft) where there are forest trees, if it is known that the water level in the ground falls below 2·13 m (7 ft) from the surface in a dry summer. The first term on the right hand side of Eq. (6·78) represents the theoretical ultimate bearing capacity of the base divided by a factor of safety of 2. In the second term, since c_s and L_e are taken at their lowest possible values, no additional safety factor is applied.

To determine the fully softened shear strength of a $1\frac{1}{2}$ in. (38·1 mm) diameter by 3 in. (76·2 mm) long specimen of the clay, the ends of the specimen are covered with celluloid discs and the curved surface wrapped in blotting paper, the paper being held in place by weak elastic bands. The covered specimen is then immersed in tap water. The soaking period has not been standardised. Ward and Green (1952) allowed soaking for 1 to 2 days; Whitaker and Cooke (1966) allowed 4 days. After soaking, the paper and end discs are carefully removed and the specimen tested in unconfined compression. Whitaker and Cooke found the average fully softened strength of London Clay from Wembley, Middlesex to be 0·0168 kgf/cm² (345 lbf/ft², 16519 N/m²) irrespective of the original strength, which was 0·977 to 2·44 kgf/cm² (2000 to 5000 lbf/ft², 95760 to 239 400 N/m²).

Skempton (1959) made a collection of data relating to bored piles in London Clay for which the ultimate bearing capacities had been established by loading tests and the strength profiles of the soils were available. He assumed $N_c = 9$ and calculated f_u from the formula $P_u = c_b N_c A_b + f_u A_s$.

Skempton examined the relationship between the value of f_u and the average cohesion $\bar{c}$ of the clay over the depth of the pile. The ratio $f_u/\bar{c}$, termed the adhesion coefficient, which Skempton designated by the symbol α, was found to vary from 0·3 to 0·6, and he suggested that $\alpha = 0.45$ was acceptable for design purposes for $\bar{c} < 2.20$ kgf/cm² (4500 lbf/ft², 215 460 N/m²) but for greater values of $\bar{c}$ the value of $f_u = 0.977$ kgf/cm² (2000 lbf/ft², 95760 N/m²) should be used. He further suggested that with these values a safety factor of 2·5 on the calculated ultimate bearing capacity should be used for single piles.

Large bored piles with enlarged bases

Large bored piles up to about 1·5 m (5 ft) in diameter and 27·5 m (90 ft) deep are formed in stiff clay by means of large mechanical augers, which have tools enabling the bases to be undercut to twice the shaft diameter or more. These foundations are sometimes called "caissons" or "deep bored cylinders". Functionally they behave as piles and will be considered as large piles. Foundations of this type carried to rock have been used for many years, but they are now used where the support is provided entirely by stiff clay.

Whitaker and Cooke (1966) investigated the behaviour of large bored piles in London Clay, both with and without enlarged bases, by placing load measuring "cells" immediately above the bases, as shown in Fig. 6.15. The difference between the load cell reading

CALCULATION OF THE ULTIMATE BEARING CAPACITY 85

and the applied load gave the load carried by frictional resistance on the shaft. The piles were constructed by a normal augering method and the concrete was placed in dry, unlined holes.

Full mobilisation of frictional resistance occurred at a settlement

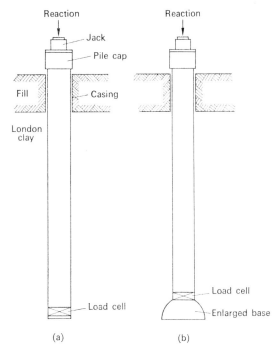

FIG. 6.15. The positions of load cells for measuring the base loads in large bored piles, (a) without base enlargement; (b) with an enlarged base.

which was between 0·5 and 1 per cent of the shaft diameter. The proportion of the frictional resistance mobilised at any given settlement was dependent on shaft diameter, but was independent of the shaft length and of whether the base was enlarged or not. The mobilisation of the base resistance increased as the settlement increased, reaching its full value at a settlement between 10 and 20 per cent of the base diameter.

The mean ultimate frictional resistance (f_u) corrected to an age of 6 months is shown in Fig. 6.16(a) plotted against mean shaft depth. The average net ultimate bearing pressure (q_u) on the base of each pile is plotted against the depth of the base in Fig. 6.16(b).

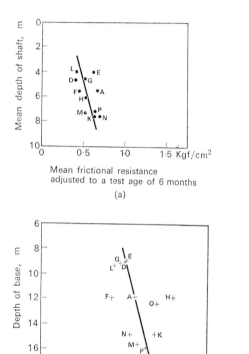

FIG. 6.16. (a) The mean ultimate frictional resistance at 6 months versus the mean depth of the shaft. (b) The mean net ultimate bearing pressure versus the depth of the base. The piles are designated by letters. The shaft diameter of piles *D, E,* and *F,* was 610 mm (2 ft) that of piles *G, L, H, A, K* and *M* was 762 mm (2 ft 6 in.) and that of piles *O, N* and *P* was 915 mm (3 ft). The bases were the same diameter as the shafts in piles *D, G, H, K* and *N* and were enlarged to twice the shaft diameter in piles *E, F, L, A, M, O* and *P*.

CALCULATION OF THE ULTIMATE BEARING CAPACITY

A normal site investigation was made, samples being taken in 101·6 mm (4 in.) diameter tubes from a borehole and the shear strengths of the clay specimens were found by undrained triaxial tests. A plot of specimen shear strength versus depth was made for a depth of 36·6 m (120 ft) and a mean line drawn through the

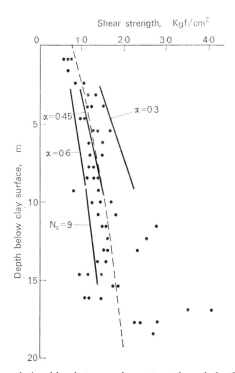

FIG. 6.17. The relationships between shear strength and depth calculated from Fig. 6.16 for $N_c = 9$ and $\alpha = 0·3$, 0·45 and 0·6, superimposed on a plot of points representing the shear strength of triaxial test specimens.

points. Figure 6.17 shows the upper portion of this plot. From the linear relationship between q_u and depth in Fig. 6.16(b) and assuming $N_c = 9$, values of c_b were calculated for the range of depths of the bases, and a line giving a relationship between these values and depth was superimposed on Fig. 6.17. The position of this line shows the actual value of c_b mobilised at the pile base at failure to be smaller than given by the mean strength resulting from triaxial tests on specimens.

The scatter of the specimen strength results in Fig. 6.17 is chiefly due to the fissured nature of the London Clay. With small triaxial test specimens, the random inclusion or exclusion of fissures from the specimens will cause a large variation in strength and the average strength will be greater than that of a specimen sufficiently large for a representative fissure pattern always to be present, as would be the case in the large volume of soil involved in shear failure at the base of the pile.

At the present time there is no recognised test for obtaining the "large scale" or "fissured" shear strength of a clay such as London Clay, although plate bearing tests have been made for the purpose (Butler, 1964). Until a test of fissured strength becomes part of common practice, the information normally available to a designer about the clay strength on a site will consist of the results of triaxial compression tests on samples taken from bore holes. Whitaker and Cooke suggest that values from the mean shear strength–depth profile should be used with a coefficient ω to give the equivalent fissured strength, so that $q_u = N_c \omega c_b$. In their investigation they obtained the value $\omega = 0.75$.

By taking values of $\bar{c}$ from the mean strength–depth profile and using values of f_u from Fig. 6.16(a), values of α were calculated. The values so obtained were found to vary little with depth, the average value being $\alpha = 0.44$. For this purpose $\bar{c}$ was calculated for different depths D from the mean strength–depth curve by the use of the expression

$$\bar{c} = \frac{\text{Area between the curve and the axis of } D}{D}$$

In a similar manner by assuming values of $\alpha = 0.3$, 0.45 and 0.6, values of $\bar{c}$ at various mean depths were calculated and lines giving these values have been superimposed on Fig. 6.17.

CALCULATION OF THE ULTIMATE BEARING CAPACITY

Should methods of determining the fissured strength become established and fissured strength values are used in design, then a value of α of about 0.6 relative to fissured strength might be appropriate.

These experiments showed that the ultimate bearing capacity of large bored piles in London Clay with or without base enlargement can be adequately expressed by the formula

$$P_u + W = \pi d_s L \alpha \bar{c} + \frac{\pi}{4} d_b^2 (N_c \omega c_b + \bar{\gamma} D) \tag{6.19}$$

For rapid design purposes the weight of the pile W is often assumed to be equal to the overburden term and the fissured nature of the clay is neglected, so that

$$P_{u\,approx} = \pi d_s L \alpha \bar{c} + \frac{\pi}{4} d_b^2 N_c c_b \tag{6.20}$$

By making this approximation the ultimate bearing capacity is overestimated by between 6 and 10 per cent for 15 m (50 ft) deep piles with enlarged bases and by between 6 and 15 per cent for 15 m (50ft) deep piles without enlarged bases, the error increasing with increase in shaft diameter and decreasing with depth.

It is generally assumed that the load-settlement characteristics and ultimate bearing capacity of a plate about 610 mm (2 ft) diameter bedded horizontally on the bottom of a borehole will be the same as those of a pile base at that depth. The plate is loaded through a column formed by a steel pile or tube. The movement of the plate under load is transmitted to dial gauges at the surface by means of push rods. The arrangement is shown in Fig. 6.18(a). The method of applying the load and measuring the settlement are the same as those used for pile loading tests and are dealt with in Chapter 9.

A method of finding shaft resistance which dispenses with the need for a load cell is to form a void by means of a collapsable container, or place a layer of compressible material on the bottom of the borehole, and then fill the shaft with concrete, as in Fig. 6.18(b). Whatever the form of the compressible body it must not take in water nor collapse under the head of wet concrete, but

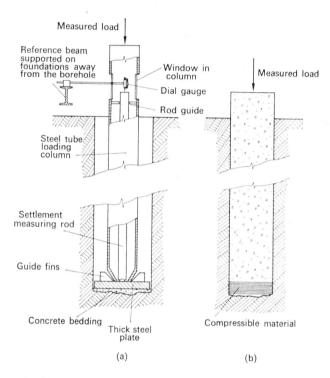

FIG. 6.18. (a) Plate bearing test at the bottom of a borehole. (b) Arrangement for determining the skin friction on a bored pile.

must offer little additional resistance to compression. A loading test gives the shaft resistance only.

In making separate plate bearing and shaft resistance tests in this manner, the two stress fields are applied independently to the soil, so that there is no interaction of one on the other such as

CALCULATION OF THE ULTIMATE BEARING CAPACITY   91

occurs in a pile. Where q_u and f_u have been determined by these means for use in practical design, it has generally been assumed that the difference from the true values is unimportant. The use of a load cell as in Fig. 6.15 ensures that the measurements of base and shaft load are made under the stress-field conditions of an actual pile.

PILE FOUNDATIONS CARRIED TO ROCK

If a pile can be driven to bedrock experience shows that, on the whole, a satisfactory foundation will result, whether it is supported on a number of normal sized piles or on large bored piles, provided the nature of the rock has been correctly assessed.

Rocks may be divided broadly into "soft" rocks such as chalk, weakly cemented sandstones, shales and mudstones, into which the lower ends of piles can be driven and which can be bored by auger rigs, and the "hard" rocks that resist pile penetration and which require normal rock drilling techniques for sinking boreholes. The ease with which piling operations can be performed and the ultimate bearing capacity of the resulting pile depend on the intrinsic strength of the rock and on the extent to which the rock mass is fissured.

A site investigation should show the slope of the rock surface and the rock itself should be sampled by core drilling. The cores will show the rock type and the lengths of core recovered in solid or shattered condition and the gaps in the recovery, together with the records of any loss of drilling water, will give some indication of the degree to which the rock is broken up by fissuring, jointing and bedding. It must be stressed that identification of the rock in the geological sense and knowledge of the strength of intact test specimens obtained from cores do not alone supply complete information for design purposes. An assessment of the degree of fissuring must be made and if possible this is best carried out by descending a shaft or borehole for visual and manual examination *in situ* of the rock blocks or fragments and of the material filling the fissures. Any inclination of the beds should also be noted.

Chalk and limestone are liable to contain solution channels and holes, which may be empty or filled with weak material. If such a condition is shown to exist, the engineer should examine the possibility of using an alternative type of foundation, e.g. a raft (it being assumed that the structure cannot be resited). If it is decided to use piles, there must be freedom to vary both their length and position as the work proceeds if holes are encountered. If large bored piles are chosen and there is no opportunity for resiting a pile should the boring strike a hole, then the bore should be continued into sound rock below the hole. If a wide joint or hole persists in the line of the pile it may be necessary to clean it out as far as possible and fill it with concrete. In such ground each pile must be treated individually, even to the extent of making small diameter drillings from the bottom of the pile borehole to prove the soundness of the underlying rock.

Driven piles

When the rock is "soft", the pile ends will penetrate more or less as into a bed of dense sand or gravel and an estimate of the bearing capacity may be obtained from penetrometer tests. If the rock is "hard", to attempt to drive the point a substantial distance into the rock would cause damage to the pile and considerable skill is required to avoid overdriving when the pile point reaches the rock surface. If the rock surface is sloping, there may be difficulty in preventing the point skidding over it, causing the pile to buckle. A method of obtaining embedment of the point used in Norway is by means of the "Oslo" point. The actual pile point is a protruding round steel bar 75—100 mm (3—4 in.) in diameter, with the lower end hollow-ground and hardened. In the case of an H section pile, the "Oslo" point is welded into a slot in the web as in Fig. 6.19 (see Bjerrum, 1957).

In many cases when piles are driven to hard rock, the ultimate bearing capacity is not governed by the support offered by the rock but by the structural strength of the pile. This will be considered in Chapter 8.

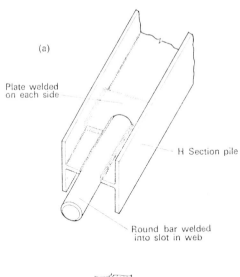

(b)

Fig. 6.19. (a) An Oslo point welded into a slot in the web of a steel bearing pile. (b) The hollow-ground tip of the Oslo point.

Bored piles

The *Civil Engineering Code of Practice No. 4* (1954), *Foundations*, gives the maximum safe bearing capacities for horizontal foundations at 2 ft (610 mm) below the rock surface under vertical static load as in Table 6.1.

If the foundation is carried more than 2 ft (610 mm) down into sound rock the bearing capacity may be increased by 20 per cent for each additional foot (305 mm) of depth up to twice the value in Table 6.1. When the rock is fissured or jointed or the beds are inclined, the Code requires a reduction in the estimated bearing capacity, to be assessed by examination of the rock *in situ*.

Since such an assessment must be based on personal experience and might be in error, it is preferable to make loading tests of the rock foundation to obtain data for design purposes.

TABLE 6.1

MAXIMUM SAFE BEARING CAPACITIES FOR HORIZONTAL FOUNDATIONS AT 2 FT (610 MM) DEPTH BELOW ROCK SURFACE UNDER VERTICAL STATIC LOADING

Type of rock	Maximum safe bearing capacity tonf/ft^2 (kgf/cm^2) (kN/m^2)		
Igneous and gneissic rocks in sound condition	100	(109·3)	(10725)
Massively-bedded limestones and hard sandstones	40	(43·8)	(4290)
Shists and slates	30	(32·8)	(3217)
Hard shales, mudstones, and soft sandstones	20	(21·9)	(2145)
Clay shales	10	(10·9)	(1072)
Hard solid chalk	6	(6·5)	(641)
Thinly-bedded limestones and sandstones	To be assessed after inspection		
Heavily-shattered rocks			

The support given by the rock will be as bearing beneath the pile base and if the rock is drilled out so that a "socket" is formed, there will be a contribution from the bond between the concrete of the shaft and the sides of the socket. The skin friction of the material overlying the rock may be positive or negative (see Chapter 8) and may be assessed either by the Dutch penetrometer or by calculation from measured shear strengths.

Bored piles up to 610 mm (2 ft) in diameter may be constructed with a socket long enough for the shaft bond to form the major contribution to bearing capacity and the value of the bond is best determined by loading tests. A downward loading rather than a pull-out test should be used. Piles of this sort are often formed by placing the foot of a steel column in the socket and casting concrete around it.

In the case of large bored piles, the load-settlement behaviour expected of the base is generally found by a plate bearing test. It will probably be found convenient to sink a cased borehole large enough for inspection of the rock to serve also for the plate bearing test. The lower end of the casing is sealed into the rock and the rock is drilled out either to the same diameter, or to a smaller diameter, to the proposed depth of the base. The base should then be cleaned of debris by hand and examined for fissuring and inclined bedding. Inspection and cleaning by hand cannot be carried out satisfactorily in a bore much less than 900 mm (about 3 ft) in diameter. A layer of concrete is then placed to overcome the irregularities of the rock and a thick steel plate bedded horizontally upon it and the concrete allowed to harden. A strong calcium sulphate plaster such as "Crystacal EN", made by British Gypsum Ltd., may be used in soft rocks, reducing the hardening time to about 2 hours.

The extent to which the shaft resistance in the rock socket is mobilised would be most correctly determined by the introduction of a load cell immediately above the base, the shaft being concreted to the level of the rock surface and the load transmitted to it by a column. Shaft resistance may, however, be determined separately by the use of a layer of compressible material at the base of the socket in the manner previously described. Whether the error, introduced into the results is important or not is not known, because no tests have been made in which a base load cell has been used, enabling shaft and base resistances to be measured simultaneously.

CHAPTER 7

THE SETTLEMENT OF SINGLE PILES AND THE CHOICE OF A FACTOR OF SAFETY

Introduction

The calculation of the settlement of a loaded pile from the known properties of the soil in which it is embedded has been attempted by a number of authors (Seed and Reese, 1955; Thurman, 1964; Coyle and Reese, 1966). The solutions so far obtained, however, are chiefly of interest to the research worker and, as yet, there is no trustworthy method of calculating the settlement of a single pile direct from soil properties that can be used for day-to-day design purposes.

The only satisfactory way of finding the amount a pile will settle under load is by making a loading test. Of course, previous experience with a particular type of pile in substantially uniform soil in a limited area enables a prediction of the settlement to be made when new piling work is undertaken, and by collecting test results, an empirical system of design rules can be evolved for the particular conditions.

It must be remembered that the settlement of a single pile, no matter how it is determined, may bear little relationship to the settlement of a foundation on a group of such piles (see Chapter 10).

If possible, however, it is best to examine a pile design in respect of both the bearing capacity and the settlement requirements and to treat them as if they were separate and unrelated, since the methods of approach to the solutions are different. The final pile

SINGLE PILES AND CHOICE OF FACTOR OF SAFETY 97

design would then be based on the requirement demanding the larger factor of safety. It commonly occurs in foundation design that a factor of safety adequate to ensure that failure of the soil does not occur is not large enough to limit the settlement to an acceptable amount.

The choice of a factor of safety to ensure that the ultimate bearing capacity is not exceeded

When the working load of a pile is obtained by computation from a driving formula using a factor of safety F then

$$\frac{\text{working load}}{\text{computed load}} = 1/F$$

But in Chapter 4 the ultimate bearing capacity by test loading, i.e. the "real" load was related to the computed load by the factor μ where

$$\frac{\text{real load}}{\text{computed load}} = \mu$$

So for any pile for which $\mu < 1/F$, the real load will be smaller than the working load. For any formula the curve given by a probability plot of μ (or $\log \mu$), as discussed in Chapter 4, can be used to develop a factor of safety for use with that formula. It is assumed that the results from which the probability curve is derived are sufficiently representative. From the probability curve is read off the value of μ (or $\log \mu$) corresponding to any chosen percentage probability of the piles having lower μ (or $\log \mu$) values. Taking $F = 1/\mu$ the value of F may be calculated. Using this factor of safety with the formula there is the probability that the chosen percentage of piles will fail at their working load. Values of F for the different formulae are given in Table 4.3 for both the $2\frac{1}{2}$ and 5% probabilities. If the bias shown by the Hiley formula is removed, as discussed in Chapter 4, F becomes 2·0 and 1·7 for the $2\frac{1}{2}$ and 5% probabilities respectively for this formula. The values in Table 4.3 were obtained from an assortment of timber, concrete and steel piles in all kinds of soil and any bias shown by the collection might not be the same as that occurring with piles of one kind only in one type of soil.

With any formula, therefore, the value of F found statistically should be treated as a partial factor, and the factor used in design should be increased to cover the possibility of the mean of results on the given site being lower than the mean of results in general, until loading tests enable the factor to be checked.

To summarise; with a "good" driving formula on an average site, using a factor of safety of 2, there is a risk of as many as 5 per cent of the piles being given working loads greater than their actual bearing capacities. Statistical theory cannot suggest which piles will be unsafe, and since if loading tests are not made there is no assurance that the conditions are up to the average, it is prudent to use a factor of safety greater than 2 in such cases. Contingencies other than the errors in using a driving formula may also need to be considered and the factor of safety increased appropriately.

Code of Practice No. 4 (1954), *Foundations*, recommends a factor of safety for average conditions of 2 for piles in non-cohesive or hard cohesive soil when the resistance is determined by driving formula only, and the pile shows no reduction of resistance on redriving. If there is reduction of resistance on redriving, the factor of safety should be $2\frac{1}{2}$ for piles in non-cohesive soil, but for piles in hard cohesive soil, a test loading should be made. If the ultimate bearing capacity is found by test loading, a safety factor of $1\frac{1}{2}$ to 2 is recommended.

Whitaker and Cooke (1966) examined the results of their tests on large bored piles statistically to give a guide to the factor of safety required in practical design based on their data. The standard deviations of f_u and q_u calculated from the regression lines shown in Figs. 6.16(a) and 6.16(b) give coefficients of variation with respect to the mean values of f_u and q_u of 16·0 and 17·9 per

SINGLE PILES AND CHOICE OF FACTOR OF SAFETY

cent respectively. These are sufficiently close for the value of 17 per cent to be taken for both.

Thus, in this case, the probability of $2\frac{1}{2}$ per cent of piles having actual ultimate bearing capacities lower than that calculated would result from the use of a "partial" factor of safety of

$$\frac{100}{100 - (2 \times 17)} = 1\cdot 5;$$

the 0·15 per cent probability limit is represented by a "partial" factor of safety of

$$\frac{100}{100 - (3 \times 17)} = 2.$$

It cannot be assumed that the same coefficient of variation would apply to another site. Differences in the construction methods and other factors peculiar to the site may effect it adversely. Further, the provision of a true margin of safety against special demands in service, contingencies in construction and other risks requires the application of an additional "partial" factor of safety.

There is no body of collected information from which a value for this second "partial" factor might be derived, and its magnitude will of necessity be governed by the designer's experience. The bearing capacity of bored piles is greatly influenced by the amount of deterioration of the clay which has occurred at the surface of the boreholes by the time the concrete is placed, and this must be allowed for in the choice of the factor of safety. The degree of disturbance due to drilling, the thoroughness with which debris has been cleared and the presence or absence of ground water in the base will all affect the proportion of the load carried by the base. The frictional resistance of the shaft will be similarly affected by drilling disturbance and by wetting, and it is probable that the longer the time the borehole remains open before concrete is placed, the greater will be any detrimental effect. If the partial factor of safety to cover these contingencies is placed at 1·5, then the overall factor of safety for use in design would be 2·25 (i.e. 1·5 × 1·5) to give a $2\frac{1}{2}$ per cent probability limit and 3 (i.e. 1·5 × 2) for a 0·15 per cent limit.

100   THE DESIGN OF PILED FOUNDATIONS

Whatever type of pile is used, the factor of safety will need to be increased if a foundation is called upon to withstand vibratory loads or large impact loads, or where the live load forms a large proportion of the total load and has to be carried for most of the time.

The choice of a factor of safety to limit the settlement at working load

When a loading test is made on a pile with the object of finding the load–settlement relationship, it must be remembered that one loading test alone tells nothing about the variation to be expected between the piles installed on the site. Differences in the behaviour of the piles can arise from the inherent variability existing between

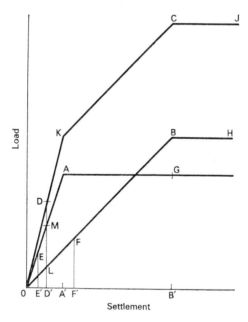

FIG. 7.1. Idealised diagrams of shaft and base resistance versus settlement and the corresponding load-settlement diagram of the pile.

piles due to small constructional differences and also from the variability of the ground from point to point. Regrettably, there are hardly any published data to guide the designer as to the variability between supposedly identical piles in uniform soil conditions. Information about the variability of the ground can be obtained by a series of penetrometer or other form of *in-situ* tests, and when this information is coupled with tests on piles at appropriate points the behaviour of the remaining piles can be estimated.

Whitaker and Cooke (1966) studied the load–settlement relationships of large bored piles on one site in London Clay and established an empirical method of design to limit the settlement of piles either with or without enlarged bases. The load–settlement diagrams fell into a general pattern which may be interpreted by reference to the behaviour of an incompressible pile in an ideally elastic–plastic soil. Thus the mobilisation of shaft and base resistance with settlement may be represented by the lines OAG and OBH respectively in Fig. 7.1 and the corresponding load–settlement curve of the pile by $OKCJ$.

The resistances R_u of the shaft and Q_u of the base become fully mobilised at A and B respectively and the ultimate bearing capacity P_u of the pile is reached at C, so that the equation

$$P_u = R_u + Q_u \qquad (7.1)$$

is represented by

$$B'C = A'A + B'B$$

If each term of the right-hand side of Eq. (7.1) is divided by the same factor of safety F, an incompatibility results, since the terms R_u/F and Q_u/F represent resistances mobilised at different values of settlement. With reference to Fig. 7.1, and taking as an example a factor $F = 3$, $P_u/3$ is represented by the ordinate $D'D$ to the load–settlement curve of the pile. The quantities $R_u/3$ and $Q_u/3$ are represented by the ordinates $E'E$ and $F'F$ to the frictional resistance and base resistance-settlement curves respectively. Numerically $D'D = E'E + F'F$, but this equation does not represent the physical state of the pile, since points D', E' and F' are not coincident. (In practice the settlement of the base would be slightly smaller than the settlement of the pile head by the amount of the shaft compression.)

It follows that for a particular settlement of the pile, say OD', which, for a working load $D'D$, will mobilise a shaft resistance $D'M$ and a base resistance $D'L$, the factors by which the ultimate bearing capacity and the ultimate shaft and base components of resistance should be divided, should all differ, so that $D'D = D'M + D'L$.

Thus $D'D = \dfrac{B'C}{F}, D'M = \dfrac{A'A}{F_s}, D'L = \dfrac{B'B}{F_b}$

and in general:

$$\frac{P_u}{F} = \frac{R_u}{F_s} + \frac{Q_u}{F_b} \qquad (7.2)$$

where F is the factor of safety applied to the ultimate bearing capacity and F_s and F_b, which are termed load factors, are applied to the ultimate frictional resistance of the shaft and the ultimate bearing resistance of the base respectively. The factors F, F_s and F_b are inter-related and of such magnitude that the individual components of resistance are calculated for the same settlement.

The expression: working load $= R_u/F_s + Q_u/F_b$ gives a simple method of design for bored piles in which the values of F_s and F_b are chosen from experience. Skempton (1966) suggested $F_s = 1\cdot5$ and $F_b = 3$ for base diameters less than 6 ft ($1\cdot8$ m).

The parameters ρ_h/d_s and ρ_b/d_b may be used to express the settlement behaviour of the shafts and bases of piles in non-dimensional terms, where

ρ_h = settlement of the pile head;
ρ_b = settlement of the pile base;
d_s = shaft diameter;
d_b = base diameter.

Also R/R_u and Q/Q_u will express the degree of mobilisation of the shaft and base resistances respectively. Any degree of mobilisation may be represented by the load factors F_s and F_b, where $F_s = R_u/R$ and $F_b = Q_u/Q$.

When considering the shaft resistance–settlement characteristics of a series of bored piles on the same site it may be assumed that the effects of soil properties and the dimensions of the piles have been eliminated by the use of the two non-dimensional parameters R/R_u and ρ_h/d_s. Thus the difference between the curves of R/R_u versus ρ_h/d_s for the individual piles may be considered to be random. Whitaker and Cooke (1966) found the mean curve for R/R_u versus ρ_h/d_s of the piles they tested and at each value of ρ_h/d_s the standard deviation (σ) of the values of R/R_u from their mean value ($\bar{r}$) was calculated. At any value of ρ_h/d_s the probabilities of a value of R/R_u being less than ($\bar{r} - \sigma$) or less than ($\bar{r} - 2\sigma$) will be 16 and $2\frac{1}{2}$ per cent respectively. Since $R/R_u = 1/F_s$, the curve ρ_h/d_s versus $1/\bar{r}$ will give the mean relationship between ρ_h/d_s and F_s, and the curve of ρ_h/d_s versus $1/(\bar{r} - \sigma)$ will give for any value of ρ_h/d_s the value of F_s which will be exceeded in 16 per cent of cases. Similarly the curve of ρ_h/d_s versus $1/(\bar{r} - 2\sigma)$ will

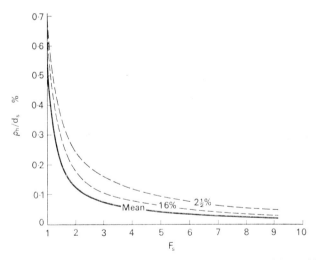

FIG. 7.2. Graph for shafts of large bored piles in London Clay with or without enlarged bases, showing the mean relationship between F_s and ρ_h/d_s and the 16 and $2\frac{1}{2}$ per cent limits of probability of variation of this relationship in experimental results.

give for any value of ρ_h/d_s the value of F_s which will be exceeded in $2\frac{1}{2}$ per cent of cases. The mean curve and the 16 and $2\frac{1}{2}$ per cent probability limits from Whitaker and Cooke's results are shown in Fig. 7.2.

Whitaker and Cooke found in their experiments that the bases of the bored piles formed by enlargement with an under-reaming tool and carefully cleared of debris by hand had resistance-settlement characteristics that were different from those of piles where no enlargement was made and no hand clearing carried out. The two types of base were therefore treated separately. In a similar manner to that used for the relationship between ρ_h/d_s and F_s a mean curve of ρ_b/d_b versus F_b for enlarged bases and curves for the 16 and $2\frac{1}{2}$ per cent probability limits were constructed and are shown in Fig. 7.3. The mean curve and the 16 per cent probability limit for bases not enlarged is given in Fig. 7.4. The value of F_b for bases not enlarged at the $2\frac{1}{2}$ per cent probability limit is approximately 30 for values of ρ_b/d_b between 0·1 and 0·5 per cent,

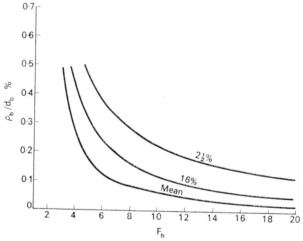

Fig. 7.3. Graph for bases in London Clay formed by enlargement with an underreaming tool and carefully cleared of debris by hand, showing the mean relationship between F_b and ρ_b/d_b and the 16 and $2\frac{1}{2}$ per cent limits of probability of variation in experimental results.

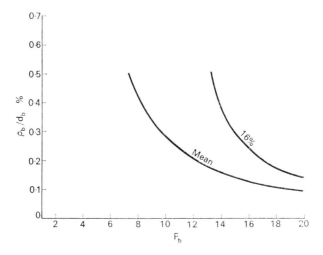

Fig. 7.4. Graph for bases in London Clay which are not enlarged and are left as cleared by the auger, showing the mean relationship between F_b and ρ_b/d_b and the 16 per cent limit of probability of variation in experimental results.

and a curve for this limit has therefore not been given. With such a high value of F_b, the contribution of the base would have little significant effect on design calculations.

To use these curves in design, the dimensions of the pile are first chosen and estimates of R_u and Q_u are made. For this purpose it is sufficiently accurate to take $R_u = \alpha \bar{c} A_s$ and $Q_u = N_c \omega c_b A_b$. It is assumed as a first approximation that the settlement of the pile base is equal to the settlement of the head and the parameters ρ_h/d_s and ρ_b/d_b appropriate to the pile dimensions and the chosen value of settlement are calculated. A value of F_s and an approximation to the value of F_b are obtained from the curves of ρ_h/d_s versus F_s and ρ_b/d_b versus F_b appropriate to the chosen probability limits. The frictional resistance and base resistance at the working load given by these values of F_s and F_b are calculated, the shaft compression estimated and a more correct value for the base settlement obtained by allowing for the shaft compression. With the new value for p_b/d_b a more correct value of F_b is found. The process of successive approximation can be repeated as necessary to give values of F_s and F_b sufficiently accurate for the purpose in hand. An average value of E for calculating the shaft compression of a bored pile is $E = 210\,000$ kgf/cm² (3×10^6 lbf/in², 20.7 kN/mm²)

The design is based on the equations

$$P_u = R_u + Q_u \qquad (7.1)$$

$$\frac{P_u}{F} = \frac{R_u}{F_s} + \frac{Q_u}{F_b} \qquad (7.2)$$

Solving these equations simultaneously

$$F = \frac{F_s F_b (1 + R_u/Q_u)}{F_b(R_u/Q_u) + F_s} \qquad (7.3)$$

Thus, the "overall" safety factor F for design can be calculated.

It is necessary to examine whether the curves of Figs. 7.2, 7.3 and 7.4 are applicable to other sites in London Clay.

The mean settlement of a base at any depth D in a weightless elastic medium is given by the equation.

$$\rho = 2(1 - v^2)I_\rho \cdot qB/E$$

where q = the average pressure on the base;
B = the width of the base;
v = Poisson's ratio of the medium;
E = Young's modulus of the medium;
I_ρ = a factor dependent on the shape, rigidity and the ratio D/B of the base.

For the case in question I_ρ may be considered to be constant. Thus it may be assumed that the immediate settlement of a circular pile base in clay is related to the pressure, diameter and soil properties by an expression of the type

$$\rho = K q d_b / E \qquad (7.4)$$

The symbol K contains the terms v and I_ρ and since it may be assumed that v will be constant for London Clay and I_ρ is constant, then K will be a constant.

Thus

$$\rho = \frac{Kq}{q_u} \cdot \frac{d_b}{E} \cdot q_u$$

where q_u is the ultimate value of q. Therefore

$$\rho = \frac{KQ}{Q_u} \cdot \frac{d_b}{E} \cdot N_c \omega c_b$$

If ρ is the settlement for a load factor F_b, so that $F_b = Q_u/Q$, then

$$\rho = \frac{K d_b N_c}{F_b E/(\omega c_b)}$$

If, therefore, for the sites under consideration the ratio $E/\omega c_b$ is constant, the settlements of bases of the same dimensions will be equal when the same load factor F_b is used. Similar reasoning may be applied to the frictional resistance of shafts of the same dimensions to show that if $E/\bar{c}$ is constant the settlements will be equal when the same load factor F_s is used. Now it has been shown by various workers that there is an approximately linear relationship between the shear strength and Young's modulus of London Clay, i.e. E/c = constant (Skempton and Henkel, 1957; Ward, Samuels and Butler, 1959). Thus, until an adequate theory is developed for the settlement of a pile relative to the soil properties, it may be taken that the immediate settlements of piles of the same size on different sites in London clay are of the same order if the same pairs of load factors F_s and F_b are used.

To cover the lack of precision in the above approach, differences in constructional techniques, and contingencies arising on the site, particularly those which affect the soil in contact with the pile, it is advisable to incorporate an additional partial factor of safety, as in the case of the bearing capacity requirement. If the constructional conditions are markedly different from those employed by Whitaker and Cooke, as for example when concrete is placed into water, then their design data will not be applicable.

If it is assumed in the case of a pile with an enlarged base that the settlement will be large enough to cause full mobilisation of the shaft resistance, then the actual settlement will depend entirely on the load–settlement characteristics of the base. If it is also assumed that the load-settlement relationships of a plate at the bottom of a borehole are the same as those of a pile base, a simpler design method in relation to settlement, such as that suggested by Burland, Butler and Dunican (1966), may be used.

Tests on stiff clays in the laboratory show that the shear strength drops for large shearing movements. Whitaker and Cooke (1966) have shown that the frictional resistance of a pile also drops from an initial peak as the settlement increases. For these reasons Burland, Butler and Dunican recommend a value of $\alpha = 0\cdot3$ be used for large settlements. Thus, the shaft resistance may be calculated on the expectation that full mobilisation will occur and by difference from the estimated working load, the base load is calculated.

A number of plate bearing tests are made on the site, covering the range of depths at which it is proposed to form the pile bases. The resulting load-settlement curves are reduced to non-dimensional form, averaged and plotted as ρ_b/d_b versus q/q_u. The plate may be loaded to failure to give the value of q_u, or q_u may be determined from the value of c_b obtained from tests on borehole samples. The ratio of the working pressure on the base to the ultimate pressure, q/q_u, is calculated and the value of ρ_b, the settlement of a base of the chosen diameter is obtained for this ratio from the plot of ρ_b/d_b versus q/q_u. A series of trial calculations may be needed to obtain a design giving an acceptable settlement.

CHAPTER 8

PILES IN SOFT SOILS

The buckling of slender piles

It has long been known that piles of normal dimensions driven through soft soil to end bearing on some strong underlying stratum do not buckle under load, provided the soil has some shear strength and is not merely liquid mud. In recent years very long and slender steel piles have been used. In one case in Norway underpinning was done with 30·5 m (100 ft) long piles made from flat bottom rail 12 cm (4·73 in.) wide and 11·6 cm (4·34 in.) high, the design load being 71 100 kgf (70 tonf, 697 kN) (Bjerrum 1957). The clay had an *in-situ* shear strength (by vane test) ranging from about 0·15 to 0·34 kgf/cm^2 (300 to 700 lbf/ft^2, 14364 to 33516 N/m^2 with a sensitivity from about 240 to 40.

For an elastic strut in an elastic medium Timoshenko (1907) showed that

$$P_{cr} = P_E \left(n^2 + \frac{B}{n^2} \right) \qquad 8.1$$

where P_{cr} = the critical or buckling load of the strut;

P_E = the Euler critical load in air;

n = the number of half sine waves in the buckled form of the strut;

L = the strut length;

k = the coefficient of lateral reaction of the medium, i.e. the resisting force per unit length of the strut per unit lateral deflection;

$B = \dfrac{kL^4}{\pi^4 EI}$

In Equation (8.1) P_{cr} is a minimum when $n = B^{\frac{1}{4}}$

The minimum value of P_{cr} is therefore

$$P_{cr} = 2P_E B^{\frac{1}{2}} = 2(kEI)^{\frac{1}{2}} \tag{8.2}$$

Thus, the buckling load is not determined by the pile length, but by the coefficient of lateral reaction of the soil and the value of EI. Model tests by Bergfelt (1957) show that Eq. (8.2) applies reasonably well for piles driven through soft clay even though the clay is not an ideally elastic medium.

If A is the cross-sectional area of the pile and p is the yield strength of the material, then if $P_{cr} > Ap$ the pile will fail by crushing and will not buckle.

Substituting for P_{cr} from Eq. (8.2) gives

$$k = p^2 A^2 / 4EI \tag{8.3}$$

For 8 in. × 8 in. × 36 lb/ft (202 mm × 202 mm × 53·6 kg/m) universal bearing piles $A^2/I = 2·67$. If the piles are of mild steel having a yield strength $p = 36,000$ lbf/in.^{2} (2530 kgf/cm^2, 0·248 kN/mm^2) and $E = 30 \times 10^6$ lbf/in.^{2} (2·11 × 10^6 kgf/cm^2, 206·8 kN/mm^2), then buckling will just occur when $k = 28·5$ lbf/in.^{2} (2·0 kgf/cm^2, 196·5 kN/m^2). If the piles are of high yield steel, so that $p = 51,500$ lbf/in.^{2} (3620 kgf/cm^2, 0·355 kN/mm^2) and $E = 30 \times 10^6$ lbf/in.^{2} (2·11 × 10^6 kgf/cm^2, 206·8 kN/mm^2), the piles will buckle when $k = 58·8$ lbf/in.^{2} (4·12 kgf/cm^2, 405·4 kN/m^2). For 10 in. × 10 in. × 57 lb/ft (254 mm × 254 mm × 84·8 kg/m) universal bearing piles the values of k at which buckling will just occur are 29 and 59·9 lbf/in.^{2} (2·04 and 4·20 kgf/cm^3, 200 and 413 kN/m^2) for mild steel and high yield strength steel respectively. Bjerrum (1957) states that for Norwegian clays with piles of typical length and width, k varies from 100 to 600 lbf/in.^{2} (7 to 42 kgf/cm^2, 689 to 4137 kN/m^2). Thus steel piles of the sections considered are not likely to buckle in the soft soils normally encountered, unless there is some initial out-of-straightness giving a predisposition towards buckling.

The buckling behaviour of an elastic pile in an elastic soil which increases in stiffness with depth was examined theoretically by Francis *et al.* (1963), who found surprisingly little difference between this and the case of a homogeneous elastic soil. In practice, however, both pile and soil are substantially elastic–plastic in behaviour, and a complete theoretical treatment of this problem is not at present available.

The derivation of k for an ideal elastic soil was made by Granholm (1929) who found

$$k = \frac{8\pi G}{2 + \log(64\pi^4 EI/B^4) - \log k} \qquad (8.4)$$

where G = the rigidity (shear modulus) of the soil;

B = the width of the pile.

Glick (1948) gave the following value for k:

$$k = \frac{1}{m_v} \cdot \frac{8\pi(1 - 2v)}{1 \cdot 13(3 - 4v)[2\log(2L/B) - 0 \cdot 443]} \qquad (8.5)$$

where m_v = the modulus of volume compressibility of the soil;

v = Poisson's ratio of the soil;

L = the length of the pile;

B = the width of the pile.

Gibson (1952) suggested that k has two values, one, k_i, applicable to rapid or instantaneous loading and the other as given by Glick which is applicable to a condition of sustained loading.

Gibson found

$$k_i = \frac{8\pi(1 - v)E_i}{1 \cdot 13(1 + v)(3 - 4v)[2\log(2L/B) - 0 \cdot 443]} \qquad (8.6)$$

where E_i = the compression modulus (Young's modulus) of the soil in a quick test such as the unconfined compression test.

Negative skin friction

In large areas of the Netherlands it is common practice to carry structures on piles driven through weak soils, that often include beds of peat, to a bearing stratum of dense sand below. A layer of sand is generally put over the natural ground to bring the surface to an acceptable level and this filling may be placed first and the piles driven through it, or it may be placed after the piles have been installed. It has been found in these conditions that a pile which shows an acceptably small settlement under a test load may settle when in use to a far greater amount than the test would suggest.

Consider a case where the soil is naturally consolidated, with the ground water level at the surface before the filling is placed and assume that installation of the pile does not disturb this condition. The pore water pressure plotted as abscissa from a vertical line AB will be represented by AC in Fig. 8.1. If a layer of

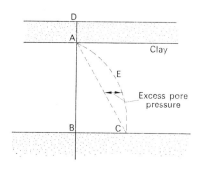

Fig. 8.1. The pore water pressure–depth relationship in a bed of clay between layers of sand, showing the excess pore pressure caused by placing the layer AD.

filling AD is now placed, the pore pressure will rise as represented by the line AEC, it being assumed that the bearing stratum below B and the fill above A are both free-draining. As this excess pore water pressure dissipates, consolidation of the weak stratum takes place and it settles.

At the place where a pile has been driven, the downward motion of the soil to the full distance is resisted by skin friction at the surface of the pile, the soil layers forming a cusp as indicated in

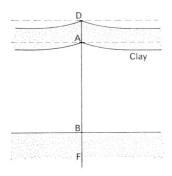

FIG. 8.2. Settlement of the soil around a pile due to consolidation.

Fig. 8.2. The downward drag of the soil on the pile is called negative skin friction, and is resisted by positive skin friction on the shaft BF in the bearing layer and by the point resistance.

Suppose a pile to have a load–settlement diagram as in Fig. 8.3, determined by test loading, which shows a settlement S for an applied load P. The broken line represents the contribution due to base resistance. The shaft resistance in the bearing stratum is included as part of the base resistance for convenience. When negative skin friction equal to P' occurs, the load $P + P'$ has to be carried on the base. This would produce a settlement S', which may well be unacceptable.

Due to negative skin friction some part of the soil weight in a zone around the pile is carried by the pile and does not contribute to overburden pressure at the level of the pile base, thus reducing the ultimate bearing resistance of the base.

In the case where the pile settles in the bearing layer due to the applied load and the load due to negative skin friction, there will be some point in the pile above which the soil moves downward relative to the pile, and below which the pile moves downward relative to the soil. Theoretical studies have been made by Buisson, Ahu and Habib (1960) of the behaviour of a single pile in these conditions.

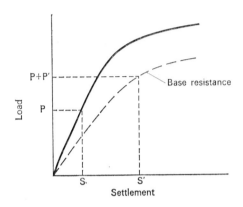

FIG. 8.3. The load–settlement diagram determined in a test and the base resistance-settlement diagram of the same pile. A test load P causes a settlement S, but with negative skin friction P' the whole load $P + P'$ is carried by the base, causing a settlement S'.

A structure is generally supported on a group of piles and the evaluation of negative skin friction in these circumstances is of more practical concern to the designer than the case of the single pile. The simplest approach to the problem of the negative skin friction affecting pile groups is that of Terzaghi and Peck (1967). If A is the area covered by a group of n piles (Fig. 8.4), S the perimeter of the group, and h_1 is the depth of filling of density γ causing consolidation in the underlying bed of clay of depth h_2 of shear strength s, then assuming that the weight of the fill over the area A is carried by the piles, the additional load per pile, Q', is given by

$$Q' = A\gamma h_1/n$$

PILES IN SOFT SOILS

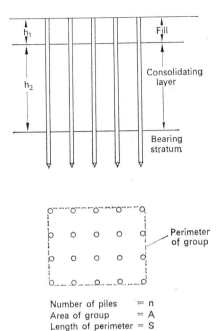

FIG. 8.4. A pile group subject to negative skin friction.

The maximum down-drag on the group due to downward movement of the bed of clay will be Sh_2s, giving a load per pile

$$Q'' = Sh_2s/n$$

If Q is the applied load per pile and Q_p the actual load carried by each pile to the bearing stratum, then, for the condition of maximum negative skin friction:

$$Q_p = Q + Q' + Q''$$

The following solution for the negative skin friction of piles in a group is due to Zeevaert (1960).

If A is the area of the group, then the number of piles per unit area is n/A. If d is the diameter of a pile and f is the ultimate skin friction per unit area at depth z, then the support given to the soil by negative skin friction per unit area per unit depth at depth z is $\pi d f n / A$.

At depth z where the overburden pressure before the piles were driven was p_0, the pressure will be now reduced to p_v because of the support given to the soil by the piles. The pressure relief at depth z is $(p_0 - p_v)$ and the rate of change of pressure relief is

$$\frac{\partial(p_0 - p_v)}{\partial z}$$

At any depth z the rate of pressure relief per unit of depth is equal to the load taken in skin friction per unit of depth, thus

$$\frac{\partial(p_0 - p_v)}{\partial z} = \frac{n\pi \, df}{A}$$

If f is proportional to p_v, so that

$$f = K p_v \tag{8.7}$$

then

$$\frac{\partial(p_0 - p_v)}{\partial z} = \frac{n\pi \, dK p_v}{A}$$

Putting $n\pi \, dK/A = m$ we have

$$\frac{\partial p_v}{\partial z} + m p_v = \frac{\partial p_0}{\partial z}$$

the solution of this equation being

$$p_v = \exp(-mx) \int \frac{\partial p_0}{\partial z} \exp(mz) \, dz + C \exp(-mz) \tag{8.8}$$

In most cases the rate of change of the initial vertical pressure with depth is equal to γ_s, the submerged density of the soil,

$$\text{i.e. } \frac{\partial p_0}{\partial z} = \gamma_s$$

PILES IN SOFT SOILS

Substituting this value and integrating Eq. (8·8) with respect to z gives

$$p_v = \exp(-mz)\left[\frac{\gamma_s}{m}\exp(mz)\right]_0^z + C\exp(-mz)$$

$$p_v = \frac{\gamma_s}{m}[1 - \exp(-mz)] + C\exp(-mz)$$

When $z = 0, p_v = 0$ so that $C = 0$

thus $p_v = \dfrac{\gamma_s}{m}[1 - \exp(-mz)]$

If h is the thickness of the bed of soil causing negative skin friction, then

$$p_h \frac{\gamma_s}{m}[1 - \exp(-mh)] \qquad (8.9)$$

The negative skin friction on one pile is

$$\int_0^h \pi\, df\, dz$$

Now $f = Kp_v$ and therefore

$$f = K\frac{\gamma_s}{m}[1 - \exp(-mz)]$$

Thus, if N.F. is the negative skin friction per pile

$$\text{N.F.} = \frac{\pi d K \gamma_s}{m}\int_0^h [1 - \exp(-mz)]dz$$

Substituting for m and integrating:

$$\text{N.F.} = \frac{A\gamma_s}{n}\left[z + \frac{\exp(-mz)}{m}\right]_0^h$$

Thus the negative skin friction is given by

$$\text{N.F.} = \frac{A}{n}(\gamma_s h - p_h) \qquad (8.10)$$

The equations

$$f = K p_v \qquad (8.7)$$

$$p_h = \frac{\gamma_s}{m}[1 - \exp(-mh)] \qquad (8.9)$$

$$\text{N.F.} = \frac{A}{n}(\gamma_s h - p_h) \qquad (8.10)$$

offer the solution for the negative skin friction per pile provided the value of K is known.

CHAPTER 9

PILE TESTING

Testing the condition of a pile after installation

The component parts of a pile, or a complete preformed pile, may be tested prior to installation. The problem to be considered here is the testing of a pile already installed in the ground.

If a driven pile hits an obstruction such as a large boulder, it may be deflected and caused to buckle or break. Any indication that this has happened may not be noticed by the driving rig operator, who would continue driving with the expectation of reaching a "set" in the normal way. With a long steel H pile it is good practice to provide a small duct down the length of the pile by welding an angle, channel or tubular section against the web as in Fig. 9.1. The straightness of the pile after installation is determined by lowering an inclinometer down the duct. A tube can be cast into a precast concrete pile for the same purpose. With a shell pile the interior can be inspected, but of course the pile is not complete until the concrete core has been placed.

In any pile formed by dropping concrete down a borehole or shell there is a risk of the concrete arching against the sides or the reinforcement. A careful check should be kept of the volume of concrete put in to ensure that the full volume of the bore is accounted for. In a bored pile that is badly made there may be a reduction of the shaft diameter by soil closing into the bore, commonly called "waisting" or "necking", or there may be gaps in the concrete of the shaft. There is no doubt that prevention is the only real solution to concreting problems and money that might otherwise be spent in testing after construction is better employed in supervision, education and good workmanship.

120   THE DESIGN OF PILED FOUNDATIONS

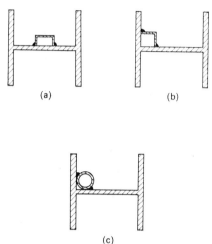

Fig. 9.1. Cross-sections of steel bearing piles with (a) channel, (b) angle and (c) tubular sections welded in place to form ducts.

The traditional ways of dealing with a doubtful pile have been to pull or dig it out, or to make a loading test upon it. These methods are direct and suitable for driven or bored piles of normal size, but with the introduction of the large bored pile, attention has been paid to other forms of testing for reasons of economy. Shaft continuity has been tested by core drilling, and the drilling record has been supplemented by inspection with a borehole camera. Attempts have been made to determine the thickness of concrete surrounding the borehole by means of apparatus consisting of a vibration transmitter with a pick-up some distance above it, lowered down a borehole. Other systems have attempted to measure the quality of the concrete by forming two boreholes in the soil close to and on opposite sides of the pile and lowering a transmitter of vibration or radiation down one and a pick-up down the other. Various methods of exciting some particular mode of vibration of the pile head and measuring the resulting response have been tried. With a freely supported prismoidal body, the time taken for a pulse to traverse the length and return after reflection at the opposite end may be used as a measure of the length. If this method is used to find the position of a gap or break in the pile, the attenuation and multiple reflection of the signal are troublesome, since the pile is not freely supported and the boundary of the concrete is not a good reflector of pulses. Thus, although the principle is attractive, no one has yet developed the method to the level of practical usefulness for normal industrial work.

The methods so far mentioned are intended to be applied to a completed bored pile that has been installed without any special provision for later tests to establish its soundness. The Centre Expérimental de Recherches et d'Études du Bâtiment et des Travaux Publics (CEBTP) in Paris first experimented on such piles with a number of methods, but in the system as finally developed for industrial use in the field, provision is made for checking the quality of the construction of bored piles on sites by forming a pair of ducts down the length of each pile to be tested during construction. This is done by attaching PVC or steel tubes of about 40 mm (about 1·6 in.) internal diameter at opposite sides of the reinforcement cage, so that the width of the pile core lies between them. The ducts are filled with water and a transmitter giving a vibration of about 12·5 c/s is lowered down one duct and a pick-up down the other. The signal transmitted across the core indicates whether the concrete is continuous or not.

It is often most expedient to make a loading test on a doubtful pile. A loading test is not, however, a proof that the construction is as specified. If "successful" it merely shows that the construction beneath the visible pile head is capable of mobilising the resistance of the ground to the required extent.

Loading tests

A loading test is made usually for one or other of the following reasons:

1. To determine the load–settlement relationship, particularly in the region of the anticipated working load.

2. To serve as a proof test to ensure that failure does not occur before a load is reached which is a selected multiple of the chosen working load. The value of the multiple is then treated as a factor of safety.
3. To determine the real ultimate bearing capacity as a check on the value calculated from dynamic or static formulae, or to obtain information that will enable other piles to be designed by empirical methods.

Very many loading tests are made for the first of these reasons and when this has been accomplished the loading may be increased and the test continued for the second or third purpose.

LOADING SYSTEMS

The following methods are used for providing the load or downward force on the pile to be tested.

(a) A platform is constructed on the head of the pile on which a mass of heavy material, which is referred to as "kentledge", is placed. This is shown diagrammatically in Fig. 9.2. The

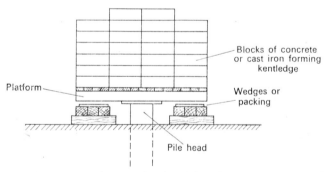

FIG. 9.2. Arrangement for carrying kentledge directly on the pile head.

load must be placed with care to obtain an axial thrust. Safety supports in the form of wedges or packings a little distance below the platform are needed on which the platform can rest to prevent the load toppling should the platform become displaced out of level as the pile settles. The method is cumbersome and is used only in exceptional cases.

(b) A bridge, carried on temporary supports, is constructed over the test pile and is loaded with a mass of heavy material. The ram of a hydraulic jack placed on the head of the pile bears on a crosshead beneath the bridge beams, so that a total reaction equal to the weight of the bridge and its load may be obtained. This is shown in Fig. 9.3.

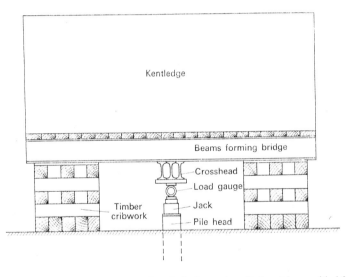

FIG. 9.3. The reaction for a hydraulic jack on the pile head is provided by kentledge supported above the pile.

(c) Anchor piles capable of withstanding an upward force are constructed on each side of the test pile, with a beam tied down to the heads of the anchor piles and spanning the test pile. A hydraulic jack on the head of the test pile obtains a reaction against the under side of the beam (Fig. 9.4). This is sometimes called the "bootstrap" method.

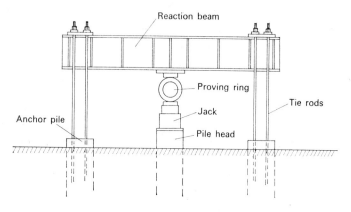

Fig. 9.4. The reaction in this arrangement is provided by anchor piles.

The supports in method (b) should be preferably more than 1·25 m (4 ft) away from the test pile, and in method (c) any anchor pile should be at least 3 test pile diameters from the test pile, centre to centre, and in no case less than 1·5 m (5 ft). When testing piles with enlarged bases the spacing should be twice the base diameter or four times the shaft diameter of the test pile, whichever is the larger, from the centre of the test pile to the centre of any anchor pile.

In method (a) the applied load is determined by weighing the platform and the material placed on it. In methods (b) and (c) it is advisable to use a proving ring or other load measuring device. If this is not possible the load on the jack ram may be calculated from the hydraulic pressure of the fluid in the jack, but if this method is used the jack with its pressure gauge should be calibrated in a testing machine under a full cycle of loading. Friction due to corrosion and wear of the jack ram and ageing of the sealing ring can cause large errors. Friction can be reduced to an acceptably low level only by maintaining the jack in good condition and by taking care to ensure that there is no eccentricity of loading, nor tilting of the surface against which the ram bears that could

PILE TESTING 125

cause the ram to bind in the bore of the jack. Corrosion is prevented and ram friction considerably reduced by chromium plating and grinding the cylindrical surface of the ram.

THE MEASUREMENT OF SETTLEMENT

The settlement of the test pile may be measured by direct levelling with a surveyor's level and staff to determine the movement of the pile head with reference to a fixed datum. It is preferable to

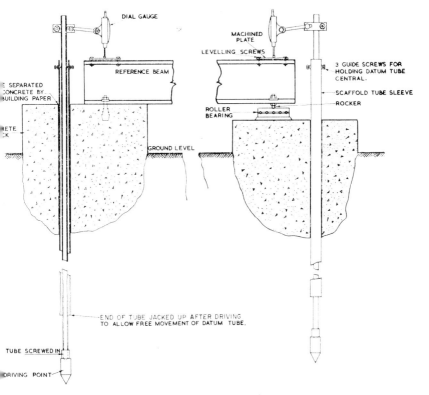

FIG. 9.5. Apparatus for measuring the movement of the ends of a beam with reference to deep datum points.

make a check to a duplicate datum. Apart from the skill of the operative, the accuracy will depend on the quality of the apparatus and the length of the sight which must be made, but a precision of reading of at least 1 mm (0·04 in.) is required.

Alternatively, the movement of the pile head may be measured by dial gauges attached to a beam that is supported on two foundations which are sufficiently far from the test pile and the reaction system to be unaffected by the ground movement. The dial gauges should read to 0·02 mm (1/1000 in.) and should be arranged in contact with a small disc of machined steel, or a piece of plate glass, bedded on the pile head or on a bracket clamped to the pile. The gauges may alternatively be fixed to the pile and make contact with suitable surfaces on the reference beam. The beam foundations may move appreciably even when 5 m (16 ft) away from the pile if the loading is by method (b), especially if the ground is soft and in such a case a deep datum point should be sunk at each end of the beam to the resistant stratum in which the pile point is embedded, as in Fig. 9.5. Movements of the beam ends are measured with reference to the datum points by dial gauges. It is of little use levelling on to the ends of the reference beam from a datum since the overall accuracy is then that of the levelling method.

Testing Procedures

The maintained load test

This is the usual method of making a pile loading test if a diagram showing the relationship between load and settlement is required. The procedure is to apply the load in stages, the load at each stage being maintained constant until the resulting settlement of the pile substantially ceases before increasing the load to the next higher stage.

Suppose a pile is to be loaded to twice the proposed working load, using a hydraulic jack and reaction system. The total load to be applied is divided into a convenient number of stages, so that the increments of load are about 25 per cent of the working load, although the first one or two increments may with advantage be larger and later ones smaller. Each increment of load is applied as smoothly and expeditiously as possible and simultaneous readings of time, and the load and settlement gauges are taken at convenient intervals as the load increases. When the load for the stage is reached, time and settlement guage readings are taken and the load is then held constant, settlement gauge readings being made at intervals of time which may be made progressively longer. A plot of settlement versus time should be made as the test proceeds and the slope of the resulting curve will give guidance as to when the load should be increased to the next stage. In the *Civil Engineering Code of Practice No. 4* (1954) a rate of movement of 0·012 in./hr (0·305 mm/hr) is taken as the limiting rate. Cooling and Packshaw (1950) recommend 0·0033 in./hr (0·084 mm/hr). A.S.T.M. D1143—57T requires a rate of settlement less than 0·012 in./hr (0·305 mm/hr) or until 2 hours have elapsed, whichever occurs first.

Usually the test proceeds with stages of loading to the maximum load, although it is not uncommon to unload the pile completely on completion of a stage at the proposed working load and hold the pile without load until the rise or "rebound" substantially ceases. The pile is then reloaded directly to the working load or to the next higher stage and the test continued. The unloading of the pile from the maximum load is carried out in stages, with a pause at each stage until the rebound substantially ceases before unloading to the next stage.

The result of a maintained load test is plotted as in Fig. 9.6, giving the curves of load and settlement versus time and of load versus the maximum settlement reached at each stage of loading. The unloading of the pile is also plotted to complete the cycle of loading.

Clearly, if the stages of loading are terminated before settlement has ceased, the true equilibrium relationship between load and settlement is not obtained. Furthermore, it often occurs in practice that the pauses are varied in length from one stage to another to suit the convenience of the operators, causing the resulting load-settlement curve to be irregular. With piles which are end bearing in sand or gravel, for which the shaft resistance is not a major component, the load-settlement curve may approach the equilibrium condition.

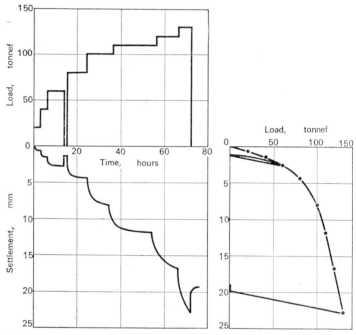

FIG. 9.6. Method of presenting the result of a maintained load test, giving plottings of load and settlement versus time and of load versus the maximum settlement at each stage of loading.

The maintained load test for determining ultimate bearing capacity.

The maintained load test is commonly used to determine the ultimate bearing capacity of a pile. An estimate of the ultimate bearing capacity must first be made so that a suitable reaction system may be provided. It is also necessary to define some physical event by which the ultimate bearing capacity, usually called "failure", can be recognised when that point is reached.

There are many definitions of failure; two well-known ones are:
(a) failing load is that which causes a settlement equal to 10 per cent of the pile diameter;
(b) failing load is that at which the rate of settlement continues undiminished without further increment of load unless this rate is so slow as to indicate that settlement may be due to consolidation of the soil.

Definition (a) is that suggested by Terzaghi (1942); definition (b) is given in *Civil Engineering Code of Practice No. 4* (1954). It is difficult to identify the point at which failure occurs when definition (b) is used, particularly with piles in cohesive soil.

The constant rate of penetration test for determining ultimate bearing capacity

This test was developed by the author for model piles (Whitaker, 1957) and it has proved equally useful for full-scale work. For convenience the test is called the C.R.P. test (Whitaker and Cooke, 1961; Whitaker, 1963).

To understand the test it is helpful to regard the pile as a device for testing the soil, and the pile movement as the means of mobilising the resisting forces. The definition of ultimate bearing capacity applicable to the test is thus:

The ultimate bearing capacity of a pile is that load at which the full resistance of the soil becomes mobilised.

In carrying out the C.R.P. test, the pile is made to penetrate the soil at a constant speed from its position as installed, while the force applied at the top of the pile to maintain the rate of penetration is continuously measured. The soil supporting the pile is stressed under conditions approaching a constant rate of strain until it fails in shear and when this occurs the ultimate bearing capacity of the pile has been reached. The test is arranged to take about the same time as an undrained shear test of a sample of the soil in the laboratory, so that the two tests have a common basis and relationships may be legitimately made between them.

It should be clearly understood that the purpose of the test is the determination of the ultimate bearing capacity of the pile, and that the force–penetration curve obtained in the test does not represent an equilibrium relationship between load and settlement, so that the settlement to be expected under working conditions is not found. Pile movement should be regarded as necessary for mobilising the forces of resistance.

The method of testing requires the generation and measurement of a downward force on the pile head, the measurement of pile movement, and some means of controlling its speed of movement. Since the force must be varied smoothly from zero to the ultimate bearing capacity, it is most convenient to use a hydraulic jack inserted between the pile head and the reaction system. The jack should have a travel greater than the sum of the final penetration of the pile and the upward movement of the reaction system. For an end-bearing pile the penetration in the test may reach 25 per cent of the pile base diameter, and for a friction pile about 10 per cent of the shaft diameter. The movement of the reaction system may be of the order of 76 mm (3 in.) if kentledge is used, and about 25 mm (1 in.) with a system of anchor piles. The pump supplying the jack may be hand or mechanically operated. For forces up to about 200 000 kgf (200 tonf), hand pumping is convenient, but if there is much "give" in the reaction system two pumps of the size normally used for a 200 000 kgf (200 tonf) jack will be required in the early stages of the test. A mechanical pump should, for preference, have an "infinitely variable" delivery, controlled either by a bleed valve or a variable speed drive. Oil cooling may be required if a bleed valve system is used. The jacking force may be measured by a load measuring device such as a proving ring, or by a pressure gauge in the jack supply line if the jack is in good condition.

The downward movement of the pile head is conveniently measured by means of a dial gauge supported on beam. In a typical C.R.P. test, a small movement of the reference beam (not exceeding 2·5 mm (0·1 in.)) is not likely to affect significantly the value obtained for the ultimate bearing capacity.

The rate of penetration may be controlled either by manually checking the time taken for successive small increments of penetration, and adjusting the pumping rate accordingly, or by attaching to the dial gauge a device similar to the "pacing ring" used to assist the maintenance of a constant rate of loading on some mechanical testing-machines. A rate of penetration of about 0·75 mm/min (0·03 in./min) has been found suitable for friction piles in clay, for which the penetration to failure is likely to be less than 25 mm. (1 in.) For end bearing piles in sand or gravel, or plate loading tests, considerably larger movements will be required to mobilise the full resistance, and rates of penetration of 1·5 mm/min (0·06 in./min) or more are required. Tests have shown that the actual rate of penetration, provided it is steady, may be half or twice these values without significantly affecting the result, and a rate should be chosen that can be held by the pumping equipment available. Major fluctuations in the rate of penetration produce corresponding undulations in the force–penetration diagram.

As the test proceeds it is advisable for the operator recording the force to make a plot of force versus penetration in order to determine when the ultimate bearing capacity has been achieved so that the test may then be terminated. If the control of the rate of penetration is manual, then it is useful to plot penetration versus time as a check on the rate of penetration achieved. This is not easily performed during the test, but the information is of use in the interpretation of undulations in the force–penetration diagram even if plotted later.

The data resulting from the test are plotted as a graph of force versus penetration. The curve in the case of a friction pile will be similar to one of those shown in Fig. 9.7(a). The force may reach a maximum value and decrease with larger penetrations, or the highest value reached may be maintained with substantially no change for two or more inches' penetration. The values of force reached at the points marked A would represent the ultimate bearing capacity in each case.

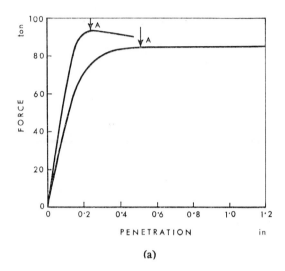

(a)

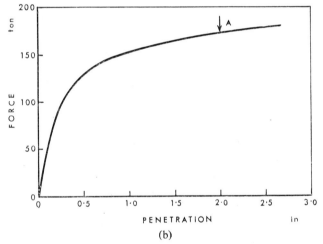

(b)

Fig. 9.7. (a) Typical force–penetration diagrams obtained with friction piles. The values of force reached at points *A* represent the ultimate bearing capacity in each case. (b) Typical force-penetration diagram obtained for an end-bearing pile.

PILE TESTING 133

The force–penetration curve in the case of an end-bearing pile will be similar to that shown in Fig. 9.7(b), and its interpretation is best made by considering its relationship to the diagram which would have been obtained had the pile been installed from the surface of the bearing stratum entirely by a constant rate of penetration technique. In Fig. 9.8, OAB represents in idealised form,

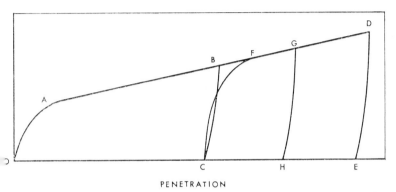

FIG. 9.8. Idealised diagram of force versus penetration for the installation of an end-bearing pile into a bearing layer and the relationship of a cycle of testing to this.

the plot of force versus penetration which would be obtained during penetration into the resistant stratum and BC during the removal of the force, with C as the final position of the pile. If the pile had been installed to a greater depth the penetration line would have extended to a point such as D, the corresponding unloaded position of the pile being E. The line OAD will be called the installation envelope.

When a C.R.P. test is made on a pile installed to C, the resulting force–penetration curve CF becomes tangential to the installation envelope at F. With further penetration the envelope would be followed to some point such as G, from which unloading would give the point H. Clearly the line CF and the point F have a unique relationship with a pile installed initially to C. F represents the point at which the resistances that are associated with the pile C are fully overcome and from which another pile of greater embedded length would be formed. Point F represents the ultimate bearing capacity of the pile installed to C. The upper part of the plot in Fig. 9.7(b) is straight or substantially straight and shows a steady increase of force with increasing penetration. A point on this line should be found corresponding to F where the unique force–penetration curve of the pile meets the installation envelope tangentially. Point A is the beginning of the substantially straight portion and the value of the force at A would represent the ultimate bearing capacity.

In practical cases, identification of the tangent point where the installation envelope is reached presents difficulty and it is usually satisfactory to take the force required to cause a penetration equal to 10 per cent of the diameter of the pile base as the ultimate bearing capacity.

CHAPTER 10

PILES IN GROUPS WITH VERTICAL LOADING

Introduction

Piles are seldom used singly; generally a group or cluster of piles is installed beneath a foundation, with a foundation slab or cap cast on the heads of the piles to distribute the load. Where the cap is cast on the ground the system will be called a piled foundation. Where the cap is formed clear of the ground it will be called a free-standing group.

The settlement of a group of vertical piles subject to a given average load per pile may be very different from the settlement of a comparable single pile under that load. Similarly, the ultimate load that can be carried by a group of piles is not necessarily the ultimate load of a single comparable pile multiplied by the number of piles in the group. This behaviour and the mechanism of interference between adjacent piles which causes it, is usually referred to as "group action". It is important in the case of friction piles in clay, not quite so important with end-bearing piles in dense sand or gravel and generally unimportant where piles are driven to rock.

The isobars of vertical stress around a pile in an elastic medium form a "bulb of pressure", approximately as shown in Fig. 10.1(a). If a number of piles of the same size are installed to form a group, the individual bulbs of pressure merge below the group to give a large bulb as indicated in Fig. 10.1(b). The widths of the loaded areas between equal isobars in the two cases are A_1 and A_2 and since in an elastic medium the settlement of a loaded area is a function of its width, the settlement of the group would be much greater than that of the single pile. Soils generally decrease in compressibility with depth so that the settlement in practical cases may be somewhat less than this simple principle would suggest.

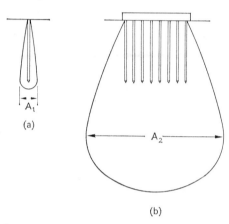

Fig. 10.1. Isobars at the same vertical stress for (a) a single pile and (b) a group of piles, showing how the width and depth of the "bulb of pressure" depend on the width of the group.

Although there have been a few attempts to devise a bearing capacity formula for pile groups that includes the interaction between adjacent piles, none has been satisfactory. The designer has therefore to make use of empirical information, chiefly obtained from tests on models, together with approximate methods and experience.

It is convenient to refer to the efficiency and the settlement ratio of groups, since these values serve as indices of the interference due to group action. By efficiency is meant the ratio of the average load per pile when failure of the group occurs to the load at failure of a comparable single pile.

There are at least two definitions of settlement ratio and care should be taken to find which is used in any instance. Whitaker (1957) defined it as the ratio between the settlement of a group to that of a comparable single pile when both carry the same fraction of their failing load. In many cases settlement ratio is taken as the ratio between the settlement of the group and a comparable single pile when the average load per pile in the group is the same as that of the single pile. With this definition, when the settlement ratio is determined for a loading representing a factor of safety F on the ultimate bearing capacity of the group, the single pile settlement would be determined at a load $\varepsilon P/F$, where P is the ultimate bearing capacity of the single pile and ε is the group efficiency. In Whitaker's definition the settlement of the single pile is taken at a load P/F. The value of the settlement ratio by Whitaker's definition is therefore the smaller when $\varepsilon < 1$.

Pile groups in clay

Where a vertical pile group supports a rigid cap on which there is a central load, each pile settles the same amount and it is

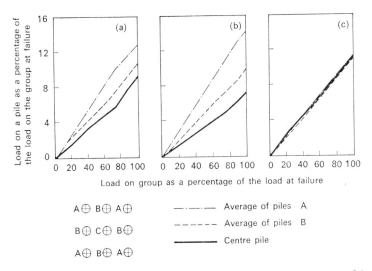

FIG. 10.2. The distribution of load among the piles in a square group of 9 model piles, for spacings of the piles centre to centre of (a) $2d$, (b) $4d$ and (c) $8d$.

commonly assumed that the piles are equally loaded. Tests show, however, that the resistances mobilised in piles at normal spacings are not equal. Whitaker (1957) measured the load carried by the piles in model free-standing groups in clay by introducing a small load gauge at the head of each pile. To present the results, the loads carried by piles occupying similar positions in the group have been averaged. In Fig. 10.2 the effect of different spacings on a square group of 9 piles is shown. At spacings equivalent to $2d$ and $4d$ (d is the pile diameter), the centre pile took the least share

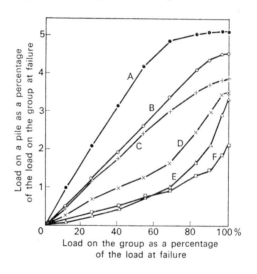

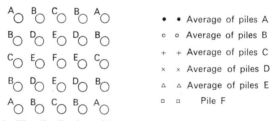

FIG. 10.3. The distribution of load among the piles in a square group of 25 model piles spaced at $2d$ centre to centre.

PILES IN GROUPS WITH VERTICAL LOADING 139

of load throughout the test, and the corner piles the greatest, but at $8d$ spacing there was no significant difference between the pile loads. The distribution of load in a 25 pile group, given in Fig. 10.3, shows that the proportion of the load taken by any pile was dependent on its distance from the centre of the group. The corner piles in this test reached their maximum load at about 80 per cent of the group failing load, and carried a constant load thereafter as the group load further increased.

Whitaker's experiments on free-standing groups in homogeneous clay demonstrated the existence of two types of failure, "block" failure and failure by individual penetration of the piles. For groups with a given length and number of piles there was a

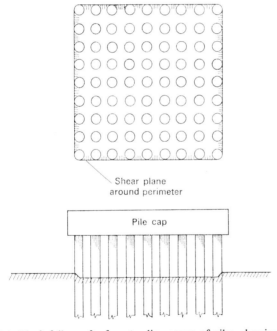

FIG. 10.4. Block failure of a free-standing group of piles, showing shearing around the perimeter and sinking of the enclosed block of soil with the piles relative to the general surface.

critical value of spacing at which the mechanism of failure changed. For spacings closer than the critical value, failure was accompanied by the formation of vertical slip planes joining the perimeter piles, the block of clay enclosed by the slip planes sinking with the piles relative to the general surface of the clay, as shown in Fig. 10.4. For wider spacings, failure was associated with local penetration of the piles into the clay. Figure 10.5 shows the efficiency versus spacing factor for square groups of free standing

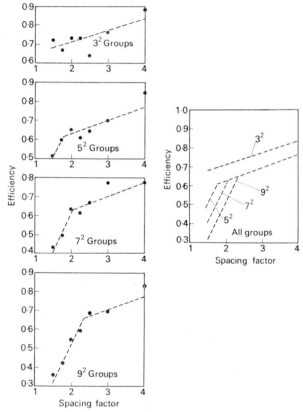

FIG. 10.5. Efficiency versus spacing factor; results of tests on model groups of free-standing piles embedded $48d$ in clay.

piles embedded $48d$ in clay. Spacing factor is the ratio of the pile spacing (centre to centre) to the pile diameter. It will be seen that the slope of each line changes abruptly at some critical value of the spacing. The more steeply inclined portions of the lines for smaller values of the spacing factor than the critical represent the condition of block failure. When a cap is cast on the surface of the soil to turn the group into a piled foundation, then, provided the cap is no larger than the perimeter of the group, in the case of a group with a spacing factor smaller than the critical value, the cap adds nothing to the efficiency. For spacings larger than the critical value, the curve of efficiency versus spacing factor for a piled foundation forms a straight line which is an extension of the line representing the condition of "block" failure, shown in Fig. 10.6, as if the cap had caused block failure in groups which would not have failed in that way had they been free-standing.

When block failure occurs in clay, the forces resisting failure are the shear strength of the clay on the shear surface forming the sides of the block and the resistance of the base of the block to bearing failure. In the case of a square group as shown in Fig. 10.4, the length of the line forming the perimeter of the group is $4[(m-1)Kd + \pi d/4]$. This is close enough to the length of the perimeter of the square enclosing the group for the latter to be taken in calculation. The area of the vertical shear surface forming the block is therefore $4L[(m-1)Kd + d]$ and the area of its base is $[(m-1)Kd + d]^2$

where d = diameter of the piles;

K = the spacing factor;

L = length of the piles;

m = number of piles in the side of the square group.

If it is assumed that the maximum resistances on the perimeter and on the base are mobilised simultaneously, then the failing load of the block will be given by the expression

$$4\bar{c}L[(m-1)Kd + d] + c_b N_c[(m-1)Kd + d]^2$$

where $\bar{c}$ = the average shear strength of the clay on the vertical surface of the block;

c_b = the shear strength at the base of the block;

N_c = the bearing capacity factor for the base of the block.

Values of N_c will vary with the $\frac{\text{depth}}{\text{width}}$ ratio of the block as shown in Table 10.1. These values are taken from results given by Skempton (1951).

TABLE 10.1

Ratio $\frac{\text{depth}}{\text{width}}$	N_c
1·5	8·4
2·0	8·6
3·0	9·1
4·0	9·3
>4·0	9·3

In Fig. 10.6 curves are given for the efficiencies of model pile groups of $3^2, 5^2, 7^2$ and 9^2 piles for different spacings calculated from the expression given above. The curves for the efficiencies of free-standing groups and piled foundations obtained experimentally in model tests in homogeneous clay have been superimposed. Since in practice the soil immediately below a pile cap is often weaker than the average strength of the bed in which the pile group is installed, the cap may not be as fully effective in producing block action as it is in the case of a homogeneous clay and the resulting ultimate bearing capacity will be lower, but not so low as would be the case for a free standing group of the same size.

Formulae for determining the efficiency of pile groups are given in some American building codes. The Converse-Labarre formula is one of these, giving the efficiency of a rectangular group of $m \times n$ piles as:

$$\varepsilon = 1 - \frac{\theta}{90}\left[\frac{(n-1)m + (m-1)n}{mn}\right]$$

PILES IN GROUPS WITH VERTICAL LOADING 143

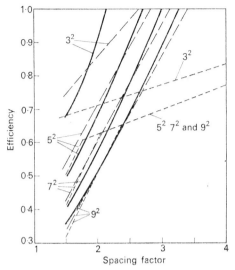

FIG. 10.6. The calculated efficiency of groups compared with the results of model tests for piles embedded 48d in clay. Dotted lines: tests on groups of free-standing piles. Broken lines: tests on piled foundations. Solid lines: as calculated for piled foundations when block failure occurs.

where ε = efficiency

m = number of rows

n = number of piles in a row

d = diameter of piles

s = spacing of piles centre to centre

$\theta = \tan^{-1} d/s$ according to the Uniform Building Code of the Pacific Coast Building Officials Conference

or $\theta = \tan^{-1} d/2s$ according to Appendix A of the Rules and Regulations of the State of California Division of Architecture.

Feld (1943) put forward a rule-of-thumb whereby the bearing value of a pile in a group is obtained by reducing its bearing capacity as a single pile by one-sixteenth for each adjacent pile. Thus, in a square group of nine piles the reduction for the corner piles is three-sixteenths, for the piles at the centres of sides five-sixteenths and for the centre pile one-half. This gives the same efficiency for the group whether the centre pile is present or not.

The most acceptable practical design method for the ultimate bearing capacity of a piled foundation in cohesive soil is to take either the value obtained by assuming block failure, or as equal to the number of piles multiplied by the bearing capacity per pile whichever is the smaller. If the group is free-standing an efficiency coefficient taken from Fig. 10.5 should be used for groups in which the spacing factor is larger than the critical value for the group, i.e. when block failure is not the limiting factor.

The suitability of a foundation on a group of friction piles is very often limited by settlement considerations. Unfortunately there are no known analytical methods of calculating either the immediate or the consolidation settlements of a pile group and recourse must be made to empirical results and to approximate methods. Whitaker (1960) found the immediate settlement ratio of model pile groups in clay for various spacing factors. The settlement ratios are given for both free standing groups and model piled foundations in Fig. 10.7 at loads equivalent to half their ultimate bearing capacities.

An approximation to the settlement resulting from consolidation caused by a group of friction piles may be found by assuming the group to behave as a pier with its base one-third the pile length up from the pile points, the area of the base being bounded by the perimeter of the group. The load is assumed to spread within the frustrum of a pyramid of side slope 30° as shown in Fig. 10.8 and to cause uniform additional vertical pressure at lower levels, the pressure being reduced in proportion to the increase in the area of the cross-section of the pyramid. The calculation for settlement then follows the normal method, assuming linear drainage in the clay (one-dimensional consolidation) between the plane and the base stratum beneath the clay.

Another approximation to the load dispersion assumes that the spreading of the load is within the frustrum of a pyramid from the

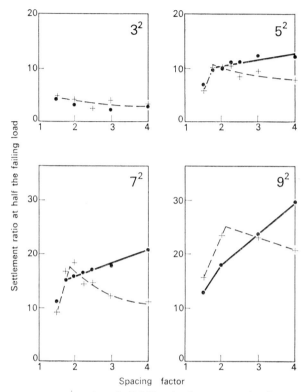

FIG. 10.7. Settlement ratio at half failing load versus spacing factor; results of test on model pile groups embedded 48d in clay. Broken lines: Free standing groups. Solid lines: Piled foundations.

top of the piles to the plane at the level of the pile tips. This represents the effect of skin friction and the slope of the side of the pyramid is chosen arbitrarily to suit the nature of the soil, varying from 12:1 for very soft clay to 4:1 for firm clay (Dunham, 1950). Below the plane of the pile tips the load dispersion is assumed to be within a pyramid having a side slope which is again chosen with regard to the type of soil, being about 2:1 for firm clay, as shown in Fig. 10.9.

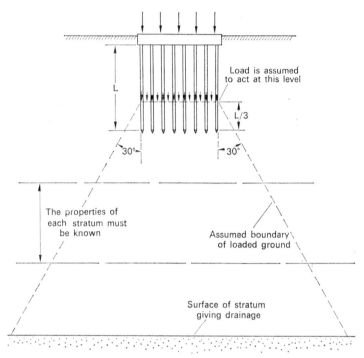

Fig. 10.8. An assumed pattern of load-spreading from a friction pile foundation for consolidation settlement calculations.

Tomlinson (1963) suggests that the load may be considered to be spread within a pyramid from the top of the piles to a plane two-thirds the depth of the piles points, the side slope of the pyramid being 4:1. For consolidation calculations the foundation is then treated as a pier standing on this larger area.

The author has no factual information as to the error involved when the settlement is calculated by any of these methods. Their essential value is that they give an approximation to the order of the settlement, enabling comparisons to be made between different designs.

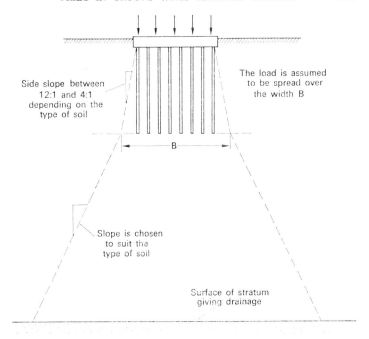

Fig. 10.9. An alternative pattern of load-spreading. The slopes of the boundaries of the loaded zone are chosen arbitrarily to outline what is assumed to be the bulb of pressure.

WHEN NOT TO USE PILES

The clay between the piles in a group of displacement piles is considerably disturbed and left in a state of stress, with an increased pore water pressure. As the pore water pressures dissipate, possibly by drainage to the piles themselves if these are of concrete or wood, the clay will consolidate and settle. Since the piles obtain their support by skin friction from this clay, the points contributing little, the piles will settle with the consolidating clay. Qualitatively, the picture is clear: where a clay shows sensitivity or marked compressibility following remoulding in laboratory testing, it is best not to construct a foundation using displacement friction piles. Quantitively, however, the position is not sufficiently well established to give numerical values to those factors which determine the conditions in which driven piles with small displacement, such as steel H piles, are suitable, or where the problem is not significant.

Research on the subject has been principally directed to showing the existence of a real problem (Cummings, Kirkhoff and Peck, 1950) but at present there is no accepted method of calculating the settlement that might occur in any given circumstances.

The displacement and heaving of clay introduces constructional problems. Where piles have to be driven into soil at the base of a cofferdam, uncontrolled lateral compression may burst the walls. To overcome this difficulty displacement piles may be started in bored holes, or steel H piles might be used, or alternatively, bored piles could be used.

Pile groups in sand

The change in the density of the sand within and around a group of displacement piles, and the consequent change in shearing resistance, form the principal features in any consideration of pile group behaviour in sand. With sands which compact when a pile is driven, the efficiency of a group of vertical free-standing piles will generally exceed 1·0 and may in certain conditions reach 3·0 or even higher. With sands that are so dense that pile driving causes loosening, an efficiency less than 1·0 may result. Model experiments have shown that "block" failure occurs in vertical free-standing groups when the pile spacing is small and that failure by individual penetration of the piles occurs when the spacing is large, in a manner similar to that found with free-standing groups in clay. Methods of calculating the ultimate bearing capacity of single displacement piles and pile groups by making allowance for the changes in the value of ϕ due to compaction or loosening of the sand have been given by Meyerhof (1959, 1960) and Kishida (1964).

The distribution of load among vertical piles in groups in sand has been examined both in full scale and in model tests and there is general agreement that the centre pile carries most load and the corner piles the least. When a driven pile is installed, before it is loaded, a residual load remains on the base which is balanced by negative skin friction on the shaft. From tests on model pile groups in which the loads carried by the base and shaft of each pile were separately measured, Walker (1964) found that if the residual load on the base of a pile was reduced to zero by the driving of adjacent piles, and it was not redriven, then that pile would carry only a small proportion of the group load irrespective of its position. Reduction of the residual base load is caused by the release of stresses between the sand grains beneath and around the pile due to disturbance, or by the pile being lifted due to the heaving of the soil when adjacent piles are driven. This points to the advisability of redriving when all the piles in the group have been installed, in order to ensure rebedding of the points and restressing of the ground.

Knowledge of the behaviour of pile groups in sand is still largely of a qualitative kind and as yet there is no coherent body of information permitting quantitative relationships to be established between the many variables involved. Thus, at present there is neither a theoretical nor an empirical approach whereby the effects of group action of piles embedded in sand can be used with assurance in design.

Piles are not often embedded entirely in sand, generally they are driven through weak material to end-bearing in a sand or gravel stratum. Provided the stratum is not underlain by weaker, compressible material the ultimate bearing capacity of the group for practical design purposes should be taken as either the ultimate bearing capacity of a single pile multiplied by the number of piles in the group, or the bearing capacity of the area enclosed by the line forming the perimeter of the group at the level of the pile points, whichever is the smaller. Both methods are conservative.

Settlement of the group can occur by compression of the material that is below the compacted zone immediately around the pile points and which remains substantially at its original density and the average standard penetration test N value of this soil should be found. The settlement of the group is not likely to exceed that of a raft foundation of the same area as the group, on soil of the same density under the same average pressure. The design may be checked by the relationship between soil pressure, width of foundation and N value for a settlement not greater than 1 in. (25·4 mm) given by Peck, Hanson and Thorburn (1953) which is reproduced in Fig. 10.10.

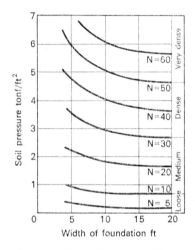

FIG. 10.10. The soil pressure that will cause a settlement of 1 in. for foundations of different widths in soils with different N values.

When a group of piles is installed to obtain end bearing in a stratum which overlies a bed of weak material as in Fig. 10.11, an investigation of both the bearing capacity and the likely consolidation settlement of the underlying material must be made. The total load should be assumed to be transferred to an area enclosed by the perimeter of the group at the pile points, with a spread of load from this area to the stratum below as indicated in Fig. 10.11. The consolidation of the weak material under this loading should be calculated following the normal methods for one-dimensional consolidation.

PILES IN GROUPS WITH VERTICAL LOADING

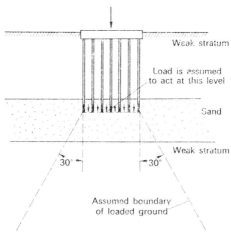

FIG. 10.11. The assumed pattern of load-spreading from an end-bearing pile foundation where the bearing stratum overlies weaker material.

If the piles are driven to rock and it is known that there are no weaker strata below, the bearing capacity should be calculated as the load per pile multiplied by the number of piles in the group. In such a case the settlement of the foundation is usually little more than the elastic shortening of the pile under load.

CHAPTER 11

HORIZONTAL FORCES ON PILES AND PILE GROUPS

Introduction

Piled foundations are often used in situations where horizontal forces have to be resisted. In most buildings these forces are small enough to be neglected, but with large buildings and bridges the resistance to wind forces, and in earthquake areas the resistance to horizontal forces caused by shocks must be adequate. In the case of a bridge the forces due to traffic acceleration, braking and turning may be important also, especially when accompanied by wind. In the case of retaining walls, quays and dolphins horizontal forces form a major part of the loading system.

The magnitude of the horizontal forces to be resisted by the foundation will be dependent upon the geometry and mass of the structure and its environment, and methods of calculating wind and earthquake forces are given in relevant Codes of Practice. The major countries liable to earthquakes have recommended design methods and the reader should consult *Earthquake resistant regulations: a world list*, 1966. The papers given in the proceedings of the three world conferences on earthquake engineering may also be of use.

In the case of a quay, horizontal forces are caused by the impact of ships during berthing, by wave impact and by wind and current pressures on ships lying against the quay, especially during a storm. The weight of the largest ship to use the berth and the velocity with which it approaches the quay must be known, so that an elastic or gravity fender may be designed that will dissipate the impact energy and limit the horizontal force on the quay. If a ship were to approach at a higher speed than anticipated, either the ship or the quay might be damaged and a design would usually be based on the assumption that the ship should suffer damage rather than the quay. If the designer has access to records of the frequency of storms and of wave heights, pressures, currents, wind speeds, etc., the design may provide resistance to the greatest possible storm, or to a storm of smaller magnitude. The frequency of occurrence of the chosen storm would be known from the records

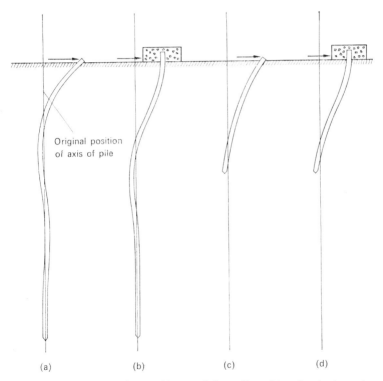

F,G. 11.1. The deflected forms of long and short piles subjected to horizontal forces at ground level. (a) a long pile with no head restraint (b) a long pile with a cap permitting no rotation of the head (c) a short pile with no head restraint (d) a short pile with a cap permitting no rotation of the head.

154 THE DESIGN OF PILED FOUNDATIONS

so that the possibility of repairs would form an assessable economic risk. The calculation of the forces on quays due to impact and storms forms a subject that is beyond the scope of this book. A summary of present knowledge has been given by Chellis (1961).

Piles are used singly for resisting horizontal forces or moments where they act as anchors, or form the foundations for transmission line poles, hoardings or bill boards. More often, however, they are used in groups in which some of the piles are inclined with the intention of providing better resistance to the applied forces. The literature dealing with the resistance of single piles and pile groups to horizontal forces is large and only a brief survey of the problem can be given here.

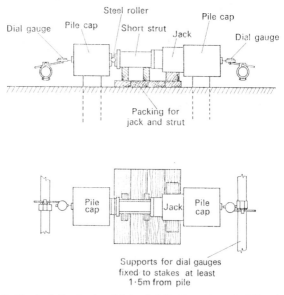

Fig. 11.2. To find the horizontal load-deflection characteristics, two piles may be jacked apart.

Single piles

When a vertical pile is deflected from its initial position by a horizontal force applied to the pile head, the deflected form of the

HORIZONTAL FORCES ON PILES AND PILE GROUPS   155

pile depends on the head conditions, the pile length, and the stiffness of both the pile and the soil. A long pile deflects as in Fig. 11.1 (a) if the head is free and as Fig. 11.1 (b) if the head is restrained by a cap so that it cannot rotate. A short pile will rotate about a point in its length as in (c) and (d).

The differential equation for the flexure of a uniform pile embedded in the soil is

$$EI\frac{d^4y}{dx^4} + p = 0 \qquad (11.1)$$

where y = the deflection of the pile at any point
x = the depth of that point from the soil surface
p = the net lateral force of the earth on the pile per unit length at that point
E = the modulus of elasticity of the pile
I = the moment of inertia of the section of the pile

In a uniform clay it is often assumed that k, the coefficient of lateral reaction (see Chapter 8) is constant, so that $p = ky$. For granular soils k is usually considered to vary linearly with depth, so that $k = n_h x$, and therefore $p = n_h xy$, where n_h is the constant of horizontal subgrade reaction, as defined by Terzaghi (1955). Palmer and Brown (1953) examined the case where the value of k varies according to the equation $k = k_L (x/L)^n$, where k_L is the value of k at the depth L, L being the pile length. They found that values of the parameter n in the range $0 < n < 1$ agreed best with test results.

The integration of Equation (11.1) can be carried out directly in simple cases, or by the method of differences for the cases where $p = k_L (x/L)^n y$, inserting the known end conditions to obtain values for the constants of integration. The result of the integration gives the deflected shape and the distribution down the pile shaft of bending moment, shearing force and the force of the earth on the pile face.

For the case where $p = ky$ and for a pile with no head restraint, such as in Fig. 11.1 (a) and with a horizontal force P applied at the ground level.

$$y = \frac{P}{2EI\beta^3} e^{-\beta x} \cos\beta x$$

$$M = -\frac{P}{\beta} e^{-\beta x} \sin\beta x$$

where M = the moment on the pile at depth x

$$\beta = (k/4EI)^{\frac{1}{4}}$$

If the head is restrained so that it cannot rotate, as in Fig. 11.1 (b) then

$$y = \frac{P}{4EI\beta^3} e^{-\beta x} (\cos\beta x + \sin\beta x)$$

$$M = \frac{P}{2\beta} e^{-\beta x} (\sin\beta x - \cos\beta x)$$

It is clear from the above equations that the curves of y and M versus x have the characteristic shape of damped waves and this is the general form in other cases for long piles. The damping in the simple cases considered above is very rapid and the wave almost disappears after one complete cycle, i.e. for values of $\beta x > 2\pi$. Thus, for pile lengths greater than $x = 2\pi/\beta = 2\pi(4EI/k)^{\frac{1}{4}}$, the length of the pile has little effect on its resistance to horizontal forces.

For further information on methods of calculating the deflection and other properties of single piles subjected to horizontal forces or head moments the reader is referred to papers by Terzaghi (1955), Broms (1964a, 1964b), Francis (1964) and Palmer and Brown (1954).

Davisson and Gill (1963) showed that in a two-layer soil system the surface layer had a controlling influence on the behaviour of a pile subjected to a horizontal force. They concluded that investigations to determine the stiffness of the soil should be most thorough within a depth equal to a few pile diameters from the ground surface.

The value assigned to the parameter k is of considerable importance. In general, k is known to vary with the type of soil, the confining pressures, the width of the pile face, the amount of deflection and the duration of loading, indeed, with all those factors which affect the resistance of a unit length of a horizontal strip foundation to a vertical load.

The application of theoretical solutions to practical design is seriously handicapped by the difficulty of obtaining the value of k and its mode of variation by the normal methods of site investigation and laboratory testing. Calculations from a knowledge of other soil parameters, as given in Chapter, 8 may be used for trial designs, but it is certainly preferable for the final design of an important structure to make a full-scale test to find the lateral deflection of the proposed type of pile in the particular soil, by installing a pair of piles and jacking their heads apart, as in Fig. 11.2. This method has been described by Wagner (1953) and others.

Groups of vertical piles

The following approximate methods are commonly used in practical design for groups of identical piles subjected to central or eccentric vertical forces or moments. The cap is assumed to be rigid and the reaction of any pile is assumed to be proportional to the displacement of the pile head.

If a vertical load V is applied at O, the centre of gravity of the pile group, the displacement of the head of each pile will be the same and the loads on the piles are therefore assumed to be equal. Thus $V = nP$, where P is the load per pile and n the number of piles.

Consider a pile group supporting a foundation that is so long that the behaviour of one row of piles forming the support of a short length of the foundation need only be examined, as in

158   THE DESIGN OF PILED FOUNDATIONS

Fig. 11.3. If an eccentric vertical force V is applied to the foundation a distance a from the centre of gravity O of the row of piles, this is equivalent to a vertical force V at O plus a moment $M = Va$ acting on the pile cap.

Considering the effect of the moment acting alone, this will cause the cap to tilt about O and the displacement of the head of

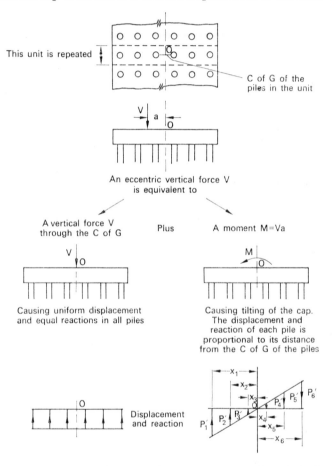

FIG. 11.3. An eccentric vertical force on a group of vertical piles.

each pile will be proportional to its distance from O, as shown in Fig. 11.3. The reaction P' of any pile due to the moment will thus be proportional to its distance from O. Thus for a row of n piles:

$$P'_1/x_1 = P'_2/x_2 = \ldots = P'_n/x_n$$

so that $\quad P'_1 = P'_1 x_1/x_1, P'_2 = P'_1 x_2/x_1, \cdots, P'_n = P'_1 x_n/x_1$

It is evident that $\quad M = P'_1 x_1 + P'_2 x_2 + \cdots + P'_n x_n$

Thus $\quad M = P'_1 x_1^2/x_1 + P'_1 x_2^2/x_1 + \ldots + P'_1 x_n^2/x_1$

Therefore $\quad P'_1 = Mx_1 \bigg/ \sum_1^n x^2$

Similarly $\quad P'_2 = Mx_2 \bigg/ \sum_1^n x^2, \ldots, P'_n = Mx_n \bigg/ \sum_1^n x^2$

Thus the total load P_1 on pile 1 due to both a vertical force through the centre of gravity and a moment is

$$P_1 = V/n \pm Mx_1 \bigg/ \sum_1^n x^2 \qquad (11.2)$$

With the moment M in the direction shown in Fig. 11.3, i.e. with the eccentric load V to the left of O, the sign of the second term on the right-hand side of Eq. (11.2) will be positive for piles to the left of the centre of gravity and negative for piles to the right. Compressive forces are counted as positive.

If a rectangular group of piles is subjected to moments about both axes XX and YY through the centre of gravity of the group, as in Fig. 11.4, as well as to a vertical force acting at the centre of gravity, then

$$P_1 = V/n \pm M_{yy}x_1 \bigg/ \sum_1^n x^2 \pm M_{xx}y_1 \bigg/ \sum_1^n y^2 \qquad (11.3)$$

The sign of the second term will be positive for piles to the left of YY and the third term will be positive for piles above XX for the moment directions as shown in Fig. 11.4.

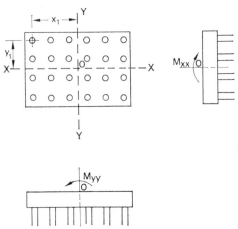

Fig. 11.4. A group of vertical piles subject to moments about two perpendicular horizontal axes.

Groups with both vertical and inclined piles

When a piled foundation has to resist a horizontal force or a moment as well as a vertical force, it is usual for some of the piles to be inclined or "battered" in order that the resultant of the external forces will be applied approximately axially to some of the piles. The calculation of the forces and moments transmitted to the head of each pile in the group presents an extremely complicated problem for which no true solution exists at present. Most approaches to the subject have been made from the direction of structural engineering, in which the piles are treated as members of a frame, the cap is assumed to be rigid and some system of forces is applied to each pile to represent the soil. In all cases there is a high order of indeterminacy and various simplifications are introduced to make a solution possible.

In the simple methods the piles are regarded as hinged at their upper ends and carry axial loads to hinges on rigid bearings at their points. In other methods the piles may be encastré at top and bottom and a lateral restraint approximating to that given by the soil may also be included. The axial displacement of the piles

HORIZONTAL FORCES ON PILES AND PILE GROUPS   161

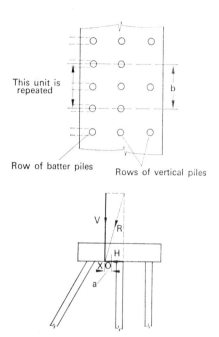

Fig. 11.5. The pile layout for a simple retaining wall foundation in which the unit of width b is repeated.

due to compression of the soil and the effect each pile has on its neighbours are ignored in all cases to enable a solution to be achieved.

An example of a simple pile group with a repeating pattern forming the foundation for a retaining wall is shown in Fig. 11.5. The foundation is of such a length that it is only necessary to determine the forces on the panel of width b.

The magnitude and line of action of the resultant of the external forces R is assumed to be known and R intersects the base of the pile cap at X, a distance a from O, the centre of gravity of the piles in the panel. V and H are the vertical and horizontal components of R at the point X.

The effect at the pile heads of a vertical force V at X is equivalent to a vertical force V at O plus a moment Va. Thus the vertical component on each of the piles can be determined by the method given previously. It is assumed that H is resisted only by the horizontal component of the axial force in the batter piles and that the vertical piles do not offer any resistance to horizontal forces. Since the vertical force and the inclination of the batter piles are known, the horizontal component required to ensure that the total force on the batter piles is axial, can now be calculated. For the foundation to be stable, the sum of the horizontal components of the batter piles must be greater or equal to H, and if this is not so the inclination of the piles or their position or number must be altered until this requirement is met.

Graphical method for piles in three directions

In the case of a foundation similar to the above, but in which the piles in a panel are inclined in not more than three directions in planes perpendicular to the length of the foundation, a graphi-

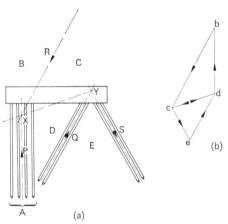

FIG. 11.6. Graphical method for piles inclined in three directions in one plane. (a) Elevation, showing the position and direction of the forces. (b) Bow's diagram for forces.

cal solution may be used. Figure 11.6 represents the elevation of the piles in a panel forming the foundation of a structure such as a quay. The cap on the piles is assumed to be rigid and all the piles and the forces on the panel are assumed to act in the same plane. The resultant R of the forces applied to the foundation is assumed to be known in magnitude, direction and position. Lines P, Q and S are drawn representing the lines of action of the rows of piles inclined in the same direction through the centres of gravity of their respective rows, e.g. P represents the line of action of the resistance of the piles A.

The direction of R meets line P in X. Lines Q and S meet in Y. Using Bow's notation as shown in Fig. 11.6 the forces on the lines XY and P are first determined and from these the forces on lines Q and S. To complete the design, it is assumed that the force in each of the three directions is equally shared among the piles inclined in that direction.

Methods based on elastic theory

One of the better known methods which provides a means of estimating pile loads when there are more than three rows, or when the piles are encastré, is that due to Vetter (1939). The method is confined to two dimensional systems. The assumptions made are:

1. The pile cap is rigid.
2. The piles are elastic.
3. The whole load is carried by the piles.
4. The resistance of a pile is concentrated at its base in the case of an end bearing pile and at one-third the length of the pile up from the base in the case of a friction pile.
5. The soil provides rigid axial bearing, with or without rigid fixity at the base or at the one-third point, but gives no other support.

The solution proceeds by assuming the pile cap to be given in turn unit horizontal, vertical and rotational displacements, the pile reactions being determined for each by elastic theory.

A modification of Vetter's method, making use of graphical methods where applicable, is included in *Civil Engineering Code of Practice No. 2* (1951), *Earth Retaining Structures*. Turzynski (1960) has given a solution based on Vetter's method, which makes use of influence coefficients, the results being obtained by the solution of a series of matrices.

Asplund (1956) has given a "Generalised elastic theory for pile groups" which allows for side resistance from an elastic soil, for piles to be encastré or hinge-ended and for piles that have any inclination in three dimensions. This method uses a matrix method of analysis to determine the axial forces in the piles.

Francis (1964) developed a method for the two-dimensional analysis of groups in which the piles could have various lengths and cross sections and have different batters and be fully or partially embedded. His method was based on the principle that a pile may be regarded as effectively fixed in the soil at some depth from its head, and he derived the positions of the points of fixity for a single pile in uniformly stiff soil and in soil which increases in stiffness with depth, and for the separate conditions of horizontal displacement and rotation of the head of the pile.

The choice of a design method

The question arises as to which method should be used for designing foundations to resist horizontal forces and in making a decision the engineer should consider the following points.

The simple methods reduce the system to an elementary determinate problem in statics by eliminating the soil and giving the piles properties that make them amenable to treatment, so that the system only barely approximates to reality. Because they do not invoke the lateral support given by the soil, the resulting designs are probably conservative.

The more complex methods, such as Vetter's and others like it, take account of the elastic properties of the piles and ignore the soil, to enable a solution to be obtained by reducing the system to an elementary indeterminate problem in statics.

Both the simple and the more complex methods are therefore on the same footing, one substituting a fictitious determinate system and the other a fictitious indeterminate one.

The methods of Asplund and Francis introduce the lateral restraint due to the soil and are therefore a step nearer reality, but they require knowledge of the parameter k, the coefficient of lateral reaction, and how k varies along the length of a pile.

All the methods so far put forward ignore the displacement of the supporting soil that accompanies the mobilisation of the axial resistance of a pile, the influence of one pile on its neighbours and the effect of the pile cap where this is cast on the ground. All methods assume that each pile acts independently, but where piles are closely spaced they do, in fact, influence each other. Both Jampel (1949) and Francis (1964) suggest that k is reduced for groups of closely spaced piles and group action may thus be just as important in any given instance for horizontal as it is for vertical forces on a pile group.

For a typical retaining wall, bridge abutment or quay design the use of a complex design method does not seem to be warranted in view of the many uncertainties that go into the problem and the questionable improvement which the complex methods can offer. If the design requires extensive use of closely spaced piles, it is advisable to make tests by forming two pile groups opposing each other, each group having the appropriate proportion of vertical and raking piles, and then jack the groups apart (see Feagin, (1953).

There are some instances, however, where a method which introduces lateral restraint due to the soil might be used with advantage, as for example, where an independent bridge pier has to be supported on relatively few long piles driven through weak material, or where the piles are free-standing, provided a realistic value of k can be obtained. The reader is advised to consult the paper by Francis (1964) in a case of this sort.

CHAPTER 12

THE DURABILITY OF PILES

PILES deteriorate due to the action of mechanical, chemical and biological agencies and if an adequate service life is to be obtained from piles in aggressive conditions a correct choice of material and its treatment are necessary.

Mechanical damage

Damage by overloading is most likely to occur when the live load forms the greater proportion of the total load and when the live load cannot be predicted with certainty. For example, free standing piles in jetties may suffer overstressing due to storms or the impact of ships, and even though the result is not immediate destruction, the damage may open the way to attack by other aggressive agencies. Reinforced concrete piles may be cracked by overstressing and although initially able to carry load, may fail later due to the reinforcement rusting because of sea water entering through the cracks.

Mechanical abrasion can be caused by sand, shingle, ice or floating debris thrown against the piles by wave action or stream flow. Wear or breakage due to the rubbing action of ships can be reduced by fitting replaceable fenders.

Chemical damage

STEEL PILES

The corrosion of steel piles is an electro-chemical phenomenon caused by potential gradients between adjacent areas of the steel surface. The steel corrodes at surfaces that are anodic to the soil and water, but is probably protected by a layer of hydrogen that is released at the cathodic surfaces. Differences in potential are caused by differences in the surface condition of the steel and by variations in the electrolyte and the amount of oxygen in solution in the water at different points in the length of a pile. The temperature and the time of exposure also determine the amount of corrosion.

The incidence of corrosion of a steel pile that is completely embedded in the ground is largely dependent on the ease with which aerated ground water can reach the pile. Thus, it is small where the permeability of the soil is low, as in a clay, but may be important in a porous soil, such as a sand, where air is present in the pores down to ground water level and dissolved oxygen may be available for some distance below. The rates of corrosion shown by experiments vary from practically nil to about 0·075 mm (0·003 in.) per year. It is a common practice to make an allowance for loss of thickness by corrosion when calculating the thickness of steel required in the wall of a tube pile or in the web and flanges of an H pile. In normal conditions that are not regarded as corrosive an increase of 1·5 mm (1/16 in.) in the thickness might be made.

For steel piles that are exposed to sea water and sea air as in the case of a jetty, the loss of steel would be least for that portion of the pile in the soil and greatest for the free standing portion. The corrosion of steel in sea water has been the subject of a number of experiments. In the tests by the Sea-Action Committee of the Institution of Civil Engineers (1920–38) the rate of loss at the surface of steel exposed to the sea water was found to vary from about 0·075 mm (0·003 in.) per year in temperate waters to about 0·175 mm (0·007 in.) per year in the tropics.

It has been noted that where the water temperature is high, the part in sea water may become so covered with marine growths that the corrosion is actually lower than that of the part in air. In temperate maritime conditions it is generally found that the zone of serious corrosion stretches from the highest point of splashing to some distance below the soil surface.

The corrosion of that portion of a pile which is embedded or permanently submerged forms a problem to be dealt with during structural design. The portion between tides and above is accessible for protective treatment and can be dealt with by appropriate maintenance. Steel piles exposed to heavy corrosive attack may be protected by a jacket of steel or concrete over the affected length. If it is found that the upper strata of the ground are corrosive, a concrete jacket may be formed by first driving a casing tube through these beds, removing the soil core and pitching and driving the pile inside the casing. When driving is completed, concrete is placed around the pile and the casing withdrawn in the normal way of forming a bored pile. A concrete jacket for the free standing portion of a pile is most conveniently constructed by the use of formwork around the pile, following normal concreting practice.

Painting will give protection if the pile can be installed without the coating being damaged by abrasion. A common form of treatment is to coat the piles with a bituminous composition or a neutralised tar. These are best applied to piles which have been thoroughly cleaned by sandblasting to bare metal, primed dry and the coating applied hot to give a thickness of about 2·5 mm (3/32 in.).

Cathodic protection of steel piles relies on the application of a potential to make the whole of the steel surface cathodic relative to the soil and water. A continuous supply of electric current must be provided between the piles and a series of anodes that are either buried in the ground nearby, or are hung in the sea beside the piles in the case of a jetty. The current may be supplied from a direct current generator, in which case the anodes might be steel scrap. Alternatively, by using a metal such as zinc or magnesium that is naturally anodic to steel in sea water, a galvanic cell may be created, the area of the anodes being made large enough to obtain the necessary current. The anodes of both systems waste away and need renewing periodically. The current required initially is in the order of 108 mA/m^2 (10 mA/ft^2) for bare steel, dropping after about 3 months to about 32 mA/m^2 (3 mA/ft^2.). When paint protection is used in addition to cathodic protection the current is considerably reduced.

Concrete piles

Any reinforced concrete structure that is exposed to sea water is liable to disintegrate completely if adequate measures are not taken to prevent this happening. The concrete itself will break down due to attack of the cement by magnesium sulphate present in the sea water if the mix is lean and pervious, but the usual cause of deterioration of reinforced concrete exposed to sea water is due to rusting of the reinforcement. This occurs if the concrete is pervious, or the thickness of the concrete cover over the reinforcement is too small, or the concrete is cracked. The rusting steel expands and forces off the cover concrete, exposing the steel to further direct corrosive attack. Durability is only obtained by adequately protecting the steel from rusting and this is best achieved by a thick layer of impermeable concrete. Thus, piles exposed to the sea must be made of rich concrete, not leaner than $1:1\frac{1}{2}:3$, well compacted to achieve the most impermeable condition and the cover over all reinforcement must be 50 mm (2 in.) or more. Concrete as rich as this would be resistant to sea water attack when well made and compacted.

Much has been said and written about the aggressive action of soils containing sulphates on concrete. There is no doubt that under laboratory conditions Portland cement concrete breaks down in solutions of sodium or magnesium sulphate, the lean mixes being more vulnerable than the rich ones. In cases where ground water containing sulphates in solution has free access to concrete and is continuously replenished so that the conditions resemble those of a laboratory immersion test, then it is probable that laboratory findings could be applied in design. In all cases the rate at which attack takes place will depend on the supply of aggressive solution to the concrete and the lower the permeabilities of the soil and the concrete, the slower will be the deterioration. In Britain the stiff clays such as the London Clay often contain sulphates, which may be of calcium, sodium or magnesium and they are thus potentially destructive to concrete. Piles installed in these clays are usually friction piles and the engineer endeavours to obtain the best possible adhesion of the clay to the concrete, so that the possibility of free ground water being present around the pile is remote and the available water is that contained in and transmitted by the capillaries and fissures. The permeability of the clay will be low, and if the concrete is rich and correctly compacted its permeability will also be low. Movement of the ground water around and towards the face of the pile is thus very restricted, furthermore, continuous entry of water into the pile must depend on the ability of the pile to get rid of water. It is frequently suggested that piles act as wicks, the water entering from the soil evaporating at the top of the foundation, but no quantitative data have been given. Such an action would cause sulphates and other salts present in the ground water to be carried into the pile. It has been suggested that water passes from a cast-in-place pile to the surrounding clay when the concrete is placed and this water will contain salts in solution from the setting concrete. The subsequent movements of solutions to or from the pile are not known, nor are the directions in which specific ions travel across the clay-concrete interface by diffusion if there is no flow of liquid. The changes in chemical composition of either the clay or the concrete that result from any ionic exchanges have not been studied.

At the present time there is no body of evidence from practice in Britain to show whether or not piles do in fact deteriorate in clays containing sulphates, what sulphate content might be tolerated, what concrete mixes should be used, or whether special cements are necessary. The need to carry out full-scale trials to give information on these questions is extremely urgent.

Because attack has been shown to occur in solutions in the laboratory, even though the conditions are not the same as in the field, a cautious engineer would consider it prudent to use a sulphate resistant cement in soil or ground water containing sulphates and its use must be regarded as a factor of safety covering risks due to ignorance of the behaviour of concrete under the particular conditions. The recommendations of the Building Research Station (1968) for foundations in ground containing sulphates should be followed until specific information is forthcoming for piles.

Biological damage

The attack of timber by fungi, usually called decay or rotting, requires moisture, air and a temperature high enough to promote fungal growth. Thus, when an untreated timber pile is fully embedded in a fairly impermeable soil, it will last indefinitely, provided it is permanently below the ground water level. Any part of a pile that is free standing, or is in soil that becomes aerated due to fluctuations in the ground water level, is liable to fungal attack.

The free standing parts of piles in marine environments are liable to attack by many burrowing organisms, the principal ones being the molluscs *Teredo* and *Bankia* and the crustaceans *Limnoria* and *Chelura*. The Sea-Action Committee of the Institution of Civil Engineers (1947) found *Limnoria* and *Chelura* to be active all round the British coast, with *Teredo* active south of the Mersey and Humber. No timber was found to be immune to attack in sea water; greenheart was most resistant, with kauri and jarrah also very resistant. Oak and softwoods were not resistant. Where the water is polluted, a condition which occurs in many harbours and estuaries, borer attack is inhibited. Greenheart in Liverpool docks was found in good condition after 60 years' service, and Danzig fir after 50 years' service in the Thames at Northfleet.

When untreated timber piles are used in tropical waters their useful life may be some months only due to borer attack. Some gum-woods have local reputations for high resistance, and protection in other cases has been given by charring the surface of the wood before installation, by sheathing with copper or by jacketing with concrete.

Creosote impregnation is generally used for protecting timber piles against fungal and borer attack. The value of the treatment depends entirely on the depth of penetration of the creosote into the wood. Various techniques are employed to obtain satisfactory penetration; e.g. vacuum is used to extract the moisture from the wood cells and pressure to force the creosote in. The success of a creosote treated timber structure also depends on the continuity of the treated layer. Thus, all untreated timber exposed by cutting or drilling must be carefully creosoted by whatever method is possible in the particular circumstances, to prevent the entry of organisms at these points into the untreated centre of the pile.

The Sea-Action Committee (1947) recommended the use of creosoted Baltic pine for British ports on the grounds of its useful economic life, although greenheart, jarrah or one of the gum-woods would be more durable. For tropical waters an efficient creosote treatment was recommended as giving high resistance to attack and ensuring a life of many years.

In many areas untreated timber piles above ground water level are attacked by termites and beetles. Preservation is usually by creosote treatment and it is important to make sure that all cuts and drill holes made on site are treated.

Damage due to industrial wastes

It may be necessary to install piles into ground contaminated by liquid wastes from industrial processes, or through filling composed of waste materials that are potentially aggressive. Harmful liquid wastes enter the soil due to spillage from plating and pickling vats, leakage from chemical plant, spillage or washings from food processing and through the dumping of liquid residues considered unsuitable for disposal as sewage. Solids such as cinders, mine waste and slags are also sources of chemicals that are aggressive when they pass into solution in ground water. In each instance where wastes are encountered the chemicals likely to be aggressive must be identified, so that the appropriate type of pile and suitable protective measures can be chosen. The help of a specialist chemist is necessary, but experience of a comparable situation is invaluable, since the rate of attack can seldom be predicted by chemistry alone.

References

AGERSCHOU, H. A. (1962) Analysis of the *Engineering News* pile formula, *J. Soil Mech. Fdns Div. Am. Soc. civ. Engrs*, **88**, SM 8, 13–34.

AMERICAN SOCIETY FOR TESTING MATERIALS (1966) *Tentative method of test for load-settlement relationship for individual piles under vertical axial load* (Designation D1143–61T) *ASTM Standards*, **11**, 358–361.

ASPLUND, S. O. (1956) Generalised elastic theory for pile groups, *Publications of the International Assoc. for Bridge and Struct. Engineering*, **16**, 1–22.

BACHUS, E. (1961) *Grundbaupraxis*, Springer, Berlin.

BARKAN, D. D. (1957) Foundation engineering and drilling by the vibration method, *Proc. 4th Int. Conf. Soil Mech., London*, **2**, 3–7.

BEGEMANN, H. K. S. Ph. (1953) Improved method of determining resistance to adhesion by sounding through a loose sleeve placed behind the cone, *Proc. 3rd. Int. Conf. Soil Mech., Zurich*, **1**, 213–217.

BEREZANTZEV, V. G. (1952) *Axial symmetrical problem of the limit equilibrium theory of earthy medium*, Gostekhizdat.

BEREZANTZEV, V. G., KHRISOFOROV, V. S. and GOLUBKOV, V. N. (1961) Load bearing capacity and deformation of piled foundations, *Proc. 5th Int. Conf. Soil Mech., Paris*, **2**, 11–15.

BERGFELT, A. (1957) The axial and lateral load bearing capacity and failure by buckling of piles in soft clay, *Proc. 4th Int. Conf. Soil Mech., London*, **2**, 8–13.

BJERRUM, L. (1957) Norwegian experience with steel piles to rock, *Geotechnique*, **7**, 73–96.

British Standards Institution (1967) *Programme for the change to the metric system in the construction industry. P D* 6030. British Standards Institution.

British Standards Institution (1967) *Methods of testing soils for civil engineering purposes*, *B.S.* 1377–1967, The British Standards Institution.

British Steel Piling Co. Ltd. (1963) *The B.S.P. Pocket Book*, The British Steel Piling Co. Ltd., London.

BROMS, B. B. (1964a) Lateral resistance of piles in cohesive soils, *J. Soil. Mech. Fdns. Div. Am. Soc. civ. Engrs*, **90**, SM2, 27–63.

BROMS, B. B. (1964b) Lateral resistance of piles in cohesionless soils, *J. Soil Mech. Fdns. Div. Am. Soc. civ. Engrs*, **90**, SM 3, 123–156.

Building Research Station (1968) *Concrete in sulphate-bearing soils and groundwaters*, Digest No. 90 (Second series) H.M.S.O. London.

BUISSON, M., AHU J. and HABIB, P. (1960) Le frottement negatif, *Annls Inst. tech. Bâtim*, **145**, 29–46.

Bureau of Reclamation (1960) *Earth Manual*, The Bureau, Denver.

BURLAND, J. B., BUTLER, F. G. and DUNICAN, P. (1966), The behaviour and design of large diameter bored piles in stiff clay, *Symposium on Large Bored Piles*, 51–71, The Institution of Civil Engineers, London.

BUTLER, F. G. (1964) Piston loading test in London Clay, *Proc. Symposium on economic use of soil testing in site investigation*, The University, Birmingham.

CASEGRANDE, A. (1932) The structure of clay and its importance in foundation engineering, *J. Boston Soc. civ. Engrs*, **14**, 168–208.

CHANDLER, R. J. (1966) Contribution to the discussion, *Symposium on Large Bored Piles*, 95–97, Inst. civ. Engrs, London.

CHANDLER, R. J. (1968) The shaft friction of piles in cohesive soils in terms of effective stress, *Civ. Eng. Pub. Wks. Review*, **63**, 48–51.

CHELLIS, R. D. (1961) *Pile foundations* (2nd edition), McGraw-Hill, New York.

CIVIL ENGINEERING CODE OF PRACTICE JOINT COMMITTEE (1951) *Civil Engineering Code of Practice No. 2* (1951) *Earth Retaining Structures*, The Institution of Structural Engineers. London.

CIVIL ENGINEERING CODE OF PRACTICE JOINT COMMITTEE (1954) *Civil Engineering Code of Practice No. 4* (1954), *Foundations*, The Institution of Civil Engineers, London.

COOLING, L. F. and PACKSHAW, S. (1950) Notes of pile-loading tests, *Chart. civ. Engr*. May, 16–20.

CORNFIELD, G. M. (1961) Simplified Hiley formula for R. C. piles, *Engineering, London*, **192**, 44.

CORNFIELD, G. M. (1964) Hiley Formula simplified without graphs, *Engineering, London*, **196**, 781.

COYLE, H. M. and REESE, L. C. (1966) Load transfer for axially loaded piles in clay, *J. Soil Mech. Fdns Div. Am. Soc. civ. Engrs*, **92**, SM2. 1–26.

CUMMINGS, A. E., KIRKHOFF, G. O. and PECK, R. B. (1950) Effect of driving piles in soft clay, *Trans. Am. Soc. civ. Engrs*, **115**, 275–285.

DAVISSON, M. T. and GILL, H. T. (1963) Laterally loaded piles in a layered soil system, *J. Soil Mech. Fdns Div. Am. Soc. civ. Engrs*, **89**, SM3, 63–94.

DE BEER, E. E. (1963) The scale effect in the transposition of the results of deep sounding tests on the ultimate bearing capacity of piles and caisson foundations, *Geotechnique*, **13**, 39–75.

DUNHAM, C. W. (1950) *Foundations of structures*, McGraw-Hill, New York.

FEAGIN, L. B. (1953) Lateral load tests on groups of battered and vertical piles. *Symposium on lateral load tests on piles.* A.S.T.M. Special Tech. Pub. No. 154, 12–20.

FELD, J. (1943) Discussion of paper by F. M. MASTERS, *Trans. Am. Soc. civ. Engrs*. **108**, 143–144.

FOREHAND, P. W. and REESE, J. L. jr. (1964) Prediction of pile capacity by the wave equation, *J. Soil Mech. and Fdns. Div Am. Soc. civ. Engrs*, **90**, SM2, 1–26.

FRANCIS, A. J. (1964) Analysis of pile groups with flexural resistance, *J. Soil Mech. and Fdns. Div. Am. Soc. civ. Engrs*, **90** SM3, 1–32.

FRANCIS, A. J., SAVORY, N. R., STEVENS, L. K. and TROLLOPE, D. H. (1963) The behaviour of slender point-bearing piles in soft soil, *Symposium on the design of high buildings*, 25–50, Hong Kong Univ. Press, Hong Kong.

REFERENCES

GARDNER, S. V. and NEW, D. H. (1961) Some experiences with prestressed concrete piles, *Proc. Instn. civ. Engrs.* **18**, 43–66, **21**, 867–91.

GIBBS, H. J. and HOLTZ, W. G. (1957) Research on determining the density of sands by spoon penetration testing, *Proc. 4th Int. Conf. Soil Mech., London*, **1**, 35–39.

GIBSON, R. E. (1950) Discussion on a paper by G. WILSON (1950) *J. Instn. civ. Engrs.* **34**, 382.

GIBSON, R. E. (1952) Reports on the stability of long piles in soft clay, Manuscript quoted by Francis *et al.* (1963)

GLANVILLE, W. H., GRIME, G., FOX, E. N. and DAVIES, W. H. (1938) *An investigation of the stresses in reinforced concrete piles during driving*, Building Research Tech. Paper No. 20, H.M.S.O., London.

GLICK, G. W. (1948) Influence of soft ground on the design of long piles, *Proc. 2nd Int. Conf. Soil Mech., Rotterdam*, **4**, 84–88.

GRANHOLM, H. (1929) *On the elastic stability of piles surrounded by a supporting medium*, Publication No. 89, Ingenirs Vetenskaps Akademien, Stockholm.

HILEY, A. (1925) A rational pile-driving formula and its application in piling practice explained, *Engineering, London*, **119**, 657 and 721.

HOLTZ, W. G. (1961) Discussion on piled foundations, *Proc. 5th Int. Conf. Soil Mech., Paris*, **3**, 254.

HOOPER, J. A. and BUTLER F. G. (1966) Some numerical results concerning the shear strength of London Clay, *Geotechnique*, **16**, 282–304.

HUIZINGA, T. K. (1951) Application of results of deep penetration tests to foundation piles, *Proc. Building Research Congress, London*, **1**, 173–179.

International Assoc. for Earthquake Engng. (1966) *Earthquake resistant regulations: a world list 1966*, Gakujutsu Bunken Fukyu-kai, Tokyo.

ISAACS, D. V. (1931) Reinforced concrete pile formulae, *Inst. Aust. Engrs. J.* **3**, 305–323.

JAMPEL, S. (1949) An analysis of groups of piles, *Concr. constr. Engng.* **44**, 201–208, 253–257.

JANBU, N. (1953) An energy analysis of pile driving using non-dimensional parameters, *Annls. Inst. tech. Bâtim.*, **63–64**, 352–360.

KERISEL, J. (1961) Deep foundations in sands: variation of ultimate bearing capacity with soil density, depth, diameter and speed of penetration, *Proc. 5th Int. Conf. Soil. Mech., Paris*, **2**, 73–84.

KISHIDA, H. (1964) *The bearing capacity of pile groups under central and eccentric loads in sands*, B.R.I. Occasional Report No. 19, Building Research Institute, Ministry of Construction, Japanese Govt.

MEYERHOF, G. G. (1951) The ultimate bearing capacity of foundations, *Geotechnique*, **2**, 301–332.

MEYERHOF, G. G. (1953) A study of the ultimate bearing capacity of piles, *Annls. Inst. tech. Bâtim.*, **63–64**, 371–373.

MEYERHOF, G. G. (1959) Compaction of sands and bearing capacity of piles, *J. Soil Mech. Fdns. Am. Soc. civ. Engrs*, **85**, SM6, 1–29.

MEYERHOF, G. G. (1960) The design of Franki piles with special reference to groups in sand, *Proc. Symposium on Pile Foundations, 6th Congress Int. Assoc. Bridge and Struct. Eng.*, Stockholm, 105–123.

REFERENCES

MEYERHOF, G. G. and MURDOCK, L. J. (1953) An investigation of the bearing capacity of some bored and driven piles in London Clay, *Geotechnique*, **3**, 267–282.

MORRISON, G. J. (1868) Contribution to the discussion of a paper by M'Alpine, W. J., *Proc. Instn. civ. Engrs*, **27**, 313.

NEIMARK, J. I. (1953) Theory of vibratory penetration, *Tech. Pap. Akad. Sci. U.S.S.R.*, **5**, 15.

PALMER, D. J. and STUART, J. G. (1957) Some observations on the Standard Penetration Test and a correlation of the test with a new penetrometer, *Proc. 4th Int. Conf. Soil Mech., London*, **1**, 231–236.

PALMER, L. A. and BROWN, P. P. (1954) Piles subject to lateral thrust; Part 2-Analysis of pressure, deflection, moment, and shear by the method of difference equations, *Supplement to Symposium on lateral load tests on piles*, A.S.T.M. Special Tech. Pub. No. 154-A, 22–44.

PALMER, L. A. and THOMPSON, J. B. (1948) The earth pressures and deflections along the embedded lengths of piles subjected to lateral thrust, *Proc. 2nd Int. Conf. Soil Mech., Rotterdam*, **5**, 156–161.

PATON, W. R. (1895) *Treatise on Civil Engineering*.

PECK, R. B. (1958) *A study of the comparative behaviour of friction piles*, Special Report No. 36, Highway Research Board, Washington, D. C.

PECK, R. B., HANSON, W. E. and THORBURN, T. H. (1953) *Foundation Engineering*, John Wiley and Sons, New York.

PRESTRESSED CONCRETE DEVELOPMENT GROUP (1964) *Draft standard specification for pretensioned prestressed piles*, The Concrete Soc., London.

ROCKEFELLER, W. C. (1967) Mechanical resonant systems in high-power applications, *Papers of Vibration Conf.*, Boston, Am. Soc. Mech. Engs.

SAURIN, B. F. (1949) The design of reinforced concrete piles with special reference to the reinforcement, *J. Instn civ. Engrs*. **32**, 80–109.

SAVINOV, O. A. and LUSKIN, A. J. (1960) *Vibratory methods of pile driving and their application in construction*, State Publishing House for Building, Architecture and Building Materials, Leningrad and Moscow.

SEA ACTION COMMITTEE OF THE INSTITUTION OF CIVIL ENGINEERS (1947) 19th Report by BRYAN, J. *Deterioration of Structures of timber, metal and concrete exposed to the action of sea water. Summary of the experimental work on timber*, The Institution of Civil Engineers London.

SEA ACTION COMMITTEE OF THE INSTITUTION OF CIVIL ENGINEERS (1960) 20th Report by LEA, F. M. and WATKINS, C. M. *The durability of reinforced concrete in sea water*, National Building Studies Research Paper No. 30. H.M.S.O. London.

SEED, H. B. and REESE, L. C. (1957) The action of soft clay along friction piles, *Trans. Am. Soc. civ. Engrs*. **122**, 731–754.

SKEMPTON, A. W. (1951) The bearing capacity of clays, *Proc. Building Research Congress, London*, **1**, 180–189.

SKEMPTON, A. W. (1959) Cast-*in-situ* bored piles in London Clay, *Geotechnique*, **9**, 153–173.

SKEMPTON, A. W. and HENKEL, D. J. (1957) Tests on London Clay from deep borings at Paddington, Victoria and South Bank, *Proc. 4th Int. Conf. Soil Mech., London*, **1**, 100–106.

SMITH, E. A. L. (1955) Impact and longitudinal wave transmission, *Trans. Am. Soc. mech. Engrs*. August, 963–973.

SMITH, E. A. L. (1962) Pile driving analysis by the wave equation, *Trans. Am. Soc. civ. Engrs*. **127**, Part 1, 1145–1171.

SMORODINOV, M. I., EROFEEV, L. V., VYAZOVIKIN, V. N. and VILLUMSEN, V. V., (1967) *Pile driving equipment*, Mashinostroenie, Moscow.

SØRENSEN, T. and HANSEN, B. (1957) Pile driving formulae—an investigation based on dimensional considerations and a statistical analysis, *Proc. 4th Int. Conf. Soil Mech.*, London, **2**, 61–65.

TAYLOR, D. W. (1948) *Fundamentals of soil mechanics*, John Wiley and Sons, New York.

TERZAGHI, K. (1942) Discussion of the Progress Report of the Committee on the Bearing Value of Pile Foundations, *Proc. Am. Soc. civ. Engrs*. **68**, 311–323.

TERZAGHI, K. (1943) *Theoretical soil mechanics*, John Wiley and Sons, New York.

TERZAGHI, K. (1955) Evaluation of coefficients of subgrade reaction, *Geotechnique*, **5**, 297–326.

TERZAGH, K. and PECK, R. B. (1967) *Soil mechanics in engineering practice*, (2nd edition) John Wiley and Sons, New York.

THURMAN, A. G. (1964) *Computed load capacity and movement of friction and end-bearing piles embedded in uniform and stratified soils*, Ph.D. thesis, Carnegie Institute of Technology.

TIMOSHENKO, S. (1907) *Bull. Polytech. Inst. St. Petersburg*.

TOMLINSON, M. J. (1957) The adhesion of piles driven in clay soils, *Proc. 4th Int. Conf. Soil Mech.*, London, **2**, 66–71.

TOMLINSON, M. J. (1963) *Foundation design and construction*, Sir Isaac Pitman and Sons, London.

TURZYNSKI, L. D. L. (1960) Groups of piles under mono-planar forces, *Struct. Engr*. **38**, 286–293.

VAN DER VEEN, C. (1953) The bearing capacity of a pile, *Proc. 3rd Int. Conf. Soil Mech.*, Zurich, **2**, 84–90.

VAN DER VEEN, C. (1957) The bearing capacity of a pile pre-determined by a cone penetration test, *Proc. 4th Int. Conf. Soil Mech,*. London, **2**, 72–75.

VETTER, C. P. (1939) Design of pile foundations, *Trans. Am. Soc. civ. Engrs*. **104**, 758–778.

WAGNER, A. A. (1953) Lateral load tests on piles for design information, *Symposium on lateral load tests on piles*, A.S.T.M. Special Tech. Pub. No. 154, 59–72.

WALKER, B. P. (1964) *Experiments on model pile foundations in sand*, Ph.D. thesis, London University.

WARD, W. H. and GREEN, H. (1952) House foundations: the short-bored pile, *Proc. Public Works and Municipal Services Congress*, 373–388.

WARD, W. H., SAMUELS, S. G. and BUTLER, M. E. (1959) Further studies of the properties of London Clay, *Geotechnique*, **9**, 33–58.

WELLINGTON, A. M. (1893) *Piles and pile driving*, Engineering News Pub. Co., New York.

REFERENCES

WEST, J. M. (1964) Further studies of foundation group behaviour in sand, Ph.D. thesis, Queen's University of Belfast.

WHITAKER, T. (1957) Experiments with model piles in groups, *Geotechnique*, **7,** 147–167.

WHITAKER, T. (1960) Some experiments on model piled foundations in clay, *Proc. Symposium on Pile Foundations, 6th Congress Int. Assoc. Bridge and Struct. Engineering, Stockholm* 124–139.

WHITAKER, T. (1963) The constant rate of penetration test for the determination of the ultimate bearing capacity of a pile, *Proc. Instn. civ. Engrs.* **26,** 119–123.

WHITAKER, T. and COOKE, R. W. (1961) A new approach to pile testing, *Proc. 5th Int. Conf. Soil Mech., Paris,* **2,** 171–176.

WHITAKER, T. and COOKE, R. W. (1966) An investigation of the shaft and base resistances of large bored piles in London Clay, *Symposium on large bored piles,* 7–49 The Institution of Civil Engineers London.

WILSON, G. (1950) The bearing capacity of screw piles and Screwcrete cylinders, *J. Inst. civ. Engrs.* **34,** 4–73.

ZEEVAERT, L. (1960) Reduction of point bearing capacity of piles because of negative friction, *Proc. First Pan-American Conf. on Soil Mech., Mexico,* **3,** 1145–1152.

INDEX

Abrasion, damage to piles by 166
Adhesion
 coefficient of 84
 on shaft 56, 65
  effect of water 83
Amplitude of vibration of pile 50
Amplitude of vibrator driver 49
Anchor or tension pile 1
Anchor pile for loading test 123
Angle of internal friction
 of soil 59–61
 from borehole samples 67
Angle of repose of soil 61
Auger rig 22
Augered pile *see* Bored pile
Arching of concrete in borehold 15

Bankia, attack on timber 171
Base
 bearing capacity of
  in clay 78–9
  in sand 66–77
  by plate test 89
 debris, clearing of 95
 debris, effect on settlement 104
Base load
 determination of 95
 measurement in large bored piles 84–5
Base resistance
 Berezantzev solution for 63
 measured value 86
 Meyerhof solution for 63–4
 mobilisation with settlement 101-5
 movement to mobilise 85
 non-cohesive soil 66
 Terzaghi solution for 61–3
Base shear zones
 comparison of pile and penetrometer 74
 dimensions of 73
Batter piles *see* Inclined piles
Bearing pile, definition of 1, 6
Bearing capacity *see also* Ultimate bearing capacity
 design considerations 96
 increase with time 90
 piles on rock 92, 93
 values on rock 93, 94
Bearing capacity factor 62
 N_c 79, 87
 N_c for group 142
Belled piles *see* Enlarged base
Bending moment in pile
 due to horizontal force 156
 due to lifting 24
Block failure of group 139
Blow-count
 record of 39
 resistant bed shown by 46
Bond of shaft in rock socket 94
Bored pile 13, 14
 borehole formation 22
 concrete placing 15
 debris in base, effect on settlement 104
 defects in 15
 design for settlement 105
 enlarged base 16
 factor of safety for 84
 normal and large 82
 softening of clay against shaft 82, 83
 test for soundness by core drilling 120
 to rock 94
 use of drilling mud 14, 15
Borehole for soil identification 7
Borer attack on timber piles 171
Boring below water table 14
Boulder clay 82

INDEX

Boulders
  breaking by chisel 14
  risk of buckling due to 119
  site investigation involving 8
British Steel Piling Co's cased pile 10
  arrangement for driving 20
Buckling
  in soft soil 109
  load 109–11
  testing for 119
Bulb of pressure of a group 135

Caisson *see* Large bored pile
Cased pile 10
Casing
  temporary 13
  use with bored pile 14
  withdrawal 15, 50
Cast-in-place pile 13
  effect of water on clay 82
Cathodic protection 168
Chelura attack on timber piles 171
Classification of piles 9
Clay
  deterioration at sides of borehole 82
  driving into sensitive 79
  friction pile in 5
  fully softened strength 83, 84
  groups in 137
  heaving, methods of avoiding 148
  proportions of shaft and base resistance of piles in 78
  remoulding around pile 79
  sensitivity may preclude use of piles 147
  stiff fissured 80
  suitability for bored piles 82
  triaxial test on 78
  unconfined compression test on 78
Cluster *see* Group
Coefficient
  adhesion 84
  for use with driving formula 45
  lateral reaction 109, 155
    calculation from soil tests 111
    importance of 110
  hammer energy losses 42
  restitution, for hammer blow 42
Compaction
  by piling 7
  comparison between pile and penetrometer 72
  effect on bearing capacity 66
  of sand by group 148
  size of zone around pile 66
Compression of shaft of bored pile 106
Compression, temporary, of soil, helmet and pile 39–41
Concrete
  arching in borehole 119
  filling to pipe pile 14
  high workability for bored piles 15
  placing by tremie 15
  in sea water 169
  shell piles 10
  sulphate attack on 169–70
Cone resistance *see* Dutch cone penetrometer
Consolidation
  negative skin friction due to 66
  settlement of group due to 144–7
Constant rate of penetration test
  equipment for 130
  general principle of 129
  interpretation of force-penetration diagram 131–4
  limitations of 130
  operational details 130, 131
  pacing dial for 131
  rate of penetration 131
  relationship with soil tests 129
  site plot of force-penetration diagram 131
Converse-Labarre formula for efficiency 142
Core drilling for testing bored piles 120
Core of soil in driven tube 14

INDEX

Corrosion
  damage to steel piles 166–8
  rate of, for steel piles 167
Corrosive soil 168
C.R.P. test *see* Constant rate of penetration test
Cushion *see* Packing

Damage to piles
  by abrasion 166
  by corrosion 166–8
  by incorrect driving 21
  by overstressing 166
Damping of vibration by soil 36, 54
Damping, values of 37
Datum for settlement measurement 124
Deep bored cylinder foundation *see* Large bored pile
Defects in bored piles 15
Diesel hammer 17
Dilation of soil by pile driving 45
Displaced soil, buoyancy due to 56
Displacement causing compaction 7
  pile, size of compaction zone 66
  piles, types of 9
Distribution of frictional force on shaft 43, 80–81
Dial gauge
  for settlement measurement 126
  with pacing ring 131
Dolly 22
Down drag *see* Negative skin friction
Drilling machine for bored piles 22
Drilling mud, use of 14
Driven pile
  breakage or buckling of 119
  rock point for 92
Driving
  compression wave in pile 24
  damage by incorrect 21
  dolly 22
  equipment for 19–22
  packing 21, 22
  pore pressure changes due to 45
  vibrators 18, 47–54

Driving formula 26–46
  application of 38–42
  as a control 45–6
  Cornfield 33
  Danish 43
  Dutch 30
  Engineering News 28
  general principles for use 46
  Hiley 32
  Janbu 31
  limitations of 42
  Morrison 29
  probability theory, application of 44
  risk with good 98
  statistical assessment 43
  takes no account of nature of soil 45
  wave equation 42
  Wellington 28
  Weisbach 31
Driving helmet 21, 22
Driving stresses 24, 25
Drop hammer 17
Dutch cone penetrometer test 7, 68
  De Beer's method for piles 74–5
  determination of pile bearing capacity by 72–6
  general principles 71
  modifications to 77
  van der Veen's method for piles 73
Dynamic formula *see* Driving formula

Earthquake forces 152
Effective stress analysis for pile 78
Efficiency formula 142
Efficiency of group, definition 136
Electricity on site 54
Empirical determination of settlement 101
End-bearing pile 5
  base resistance of 66
  constant rate of penetration test on 132
  in groups 135

End-bearing pile (*cont.*)
  on rock 91
Energy loss in steam hammer 38
Engineering News formula 43
Enlarged base
  by boring methods 16
  by Franki system 13
  on large bored pile 84
Equipment for pile installation 17
Equivalent diameter of pile base 73
Eytelwein's formula 30, 43

Factor of safety
  bored piles
    for bearing failure 84, 98–100
    for limiting settlement 100–8
  code of practice recommendation 98
  driving formula, statistical determination of 97
  separate values required for shaft and base 101–5
Failure load of pile, definition of 129
Failure mechanism at base 62–4
Feld, reduction rule for group 143
Fissured clay, variation of strength results 88
Fissured rock 91, 95
Fissured strength of clay, relation to average triaxial strength 88
Flexure due to horizontal force 155
Formula *see* Driving formula; Static formula
Formula, loading test supplies check on 122
Franki pile 13
Free-standing group *see* Group, free standing
Frequency of vibrator driver 49
Friction pile 6
  constant rate of penetration test on 131
  in groups 135
Frictional resistance of soil in vibrator driving 50

Fully softened shear strength of clay 83–4
Fungal attack on timber piles 171

Gaps in concrete piles 15
Gauges, dial, for settlement measurement 126
Grab, use in bored pile 14
Gravel, base resistance in 66
  inflow into borehole 15
  soil tests in 8
Ground conditions *see also* Site investigations 7
  installation method determined by 9
  variability by *in-situ* testing 101
  where piles are used 5
Ground water containing sulphates effect on concrete 169–70
Group
  above bed of weak soil 150
  action 135
  bearing capacity formula not available for 136
  block failure 139
    in piled foundation 141
  bulb of pressure 135
    in clay 137
  consolidation settlement 144–7
  Converse-Labarre efficiency formula for 142
  critical value of spacing for block failure 140
  definitions 135
  design for settlement 144–7
  distribution of load between piles in 138, 148, 157–65
  driven to rock 151
  eccentric vertical force on 157–60
  efficiency, definition 136
  end-bearing piles in 135, 149
  Feld's reduction rule 143
  forces in piles due to external moment 157–60
  free-standing 135
  friction piles in 135
  horizontal force on 160

INDEX

Group (*cont.*)
  inclined forces on 162–3
  inclined and vertical piles in 160
  model tests 136
  piled foundation, calculation of ultimate bearing capacity 141
  practical design for bearing capacity 144
  in sand 148–9
    block failure of 148
    loss of bearing capacity due to heaving 149
  settlement
    influence on design 144
    not same as single pile 135
    ratio, definition 136
  spacing
    criticil value of 140
    of piles 138–43

H piles, advantages of 10
  choice of weight 22
  comparison of drop with diesel 39
  drop 17
  internal 10
  power 17
  rated energy 39
  vibratory impact 55
Hammer coefficient 42
Hanging leaders 20
Heaving of soil 148
  assumed in foundation solution 62
  loss of bearing capacity due to 149
Helmet 21, 22
Hiley's formula 43
Holmpress pile 13
Horizontal force
  bending moment due to 156
  coefficient of lateral reaction k 155
  flexure pile equation for 155
  on group 160–5
  head fixity, effect of 154
  importance of value of k 165
  layered soil 156
  length of pile, effect of 156
  loading test 157
  maritime works 152–3
  methods of calculation, assessment of 165
  occurrence 152, 154
  practical design difficulties 157
Hydraulic jack for loading test 123

Impact driver, vibratory type 54
Impact
  laws of 30
  incorrect application in formulae 42
Inclined bedding in rock 95
Inclined piles 160
Inclinometer test for buckled piles 119
Incremental load test *see* Maintained load test
Industrial wastes, attack on piles 172
In-situ tests 8
Inspection of rock in borehole 15, 94
Installation of piles, methods of 9, 17
Integrity, structural *see* Soundness
Internal friction, angle of 59–61, 67
Internal hammer 10

Janbu's formula 43
Jetting 21
  as aid to vibrator driving 53

Kentledge for loading test 122–3

Large bored pile
  calculation of ultimate bearing capacity 89
  settlement in London clay 101
Lateral force *see* Horizontal force
Lateral reaction, coefficient of 111
Lateral reinforcement 24
Leaders for pile driving 19
Length of piles 5

Lifting piles 23, 24
Lifting points, position of 23, 24
Limit equilibrium, in resistance calculation 63
Limnoria, attack on timber piles 171
Lining, temporary 13
Load application, methods of 122–4
Load-cell for base load measurement 95
Load factor on shaft and base 102
Load measurement in loading test 124
Load-settlement relationship
  diagrams for shaft and base 101
  loading test for 121
  maintained load test, plot of result 127
  use of non-dimensional plot 103–8
Loading test
  distance of anchors from test pile 124
  experiments on large bored piles 84–7
  kentledge for 122–3
  limitations of one test 100
  load measurement 124
  maintained load test procedure 126
  movement of reaction system 130
  purpose of 121
  rebound in 127
  settlement measurement 124
  where prudent to make 45
London clay
  bored pile experiments in 84
  fissured strength 88
  fully softened shear strength of 84
  settlement of bases in 106, 107
  sulphates in 169
Loose sand, compaction by piling 7

Magnesium sulphate, effect on concrete 169
Maintained load test
  limiting rate of settlement 126
  practical limitations 127
  procedure 126
  size of load increment 127
Mandrel driven piles 10
Maritime works, screw piles in 17
Mechanics applied to foundation design 2
Meyerhof's formula 65
Model pile, distribution of skin friction 81
Model tests on groups 136

Negative skin friction
  causes 112
  group
    Tarzaghi and Peck's solution 114–15
    Zeevaert's solution 115–18
  on pile driven to rock 94
  overloading of pile point 113
Newton's laws of impact 30
  incorrect application of 42
Non-cohesive soil
  base resistance in 66
  boring in 14, 15
Non-dimensional plot of load-settlement 103–5, 108
Non-displacement pile 9, 13

Oslo point 92

Packing 21, 22
Paint, protection of steel piles by 168
Partial factor of safety 107
Paton's formula 59
Penetrometer *see also* Dutch cone penetrometer test; Standard penetration test
  in-situ testing by 67
  Dutch cone 68
  dynamic 68
  Standard Penetration Test 68
  static 68
  types of 77

INDEX

Pile
 anchor *see* Anchor pile
 bored 13, 14
 "cased" 10
 cast-in-place 10
 classification 9
 definition of 1
 displacement 9
 driven cast-in-place 11
 end-bearing *see* End-bearing pile
 Franki 13
 friction *see* Friction pile
 H section 10
 handling and lifting 23, 24
 Holmpress 13
 installation equipment 17
 mandrel driven 10
 non-displacement 9, 13
 pipe 13, 14
 point-bearing *see* End-bearing
 precast concrete 10
 proprietary systems 11
 Raymond 10
 reinforcement in 22, 23
 screw 16
 shell
  concrete 10
  steel 10
 slender 109
 structural strength of 22
 tension *see* Tension pile
 timber *see* Timber pile
 treatment of doubtful 120
 tube open at base 13
 tubular 10
 Vibro 13
 West 10
 when not to use 147
 where used 5
Pile driving equipment 19–22
Piled foundation, where used 5
 see also Groups, piled foundation
Pipe pile 13, 14
Plant for piling 17
Plate bearing test
 in borehole 89
 bedding of plate for 95
 load-settlement plot 108
 on rock 94
Plug
 of concrete for driving tube 12
 of soil in driven tube 14
Pore pressure
 changes due to pile driving 45
 dissipation causing take up 79–80
 due to soil stresses 78
 on pile, lack of experimental data 58
Pore water at surface of borehole 82
Power hammer 17
Principal stresses at base of pile 59–60
Probability
 factor of safety by 97
 theory applied to driving formulae 44
Proof tests 122
Pull-out tests 94

Quake of ground 37
 on driving 35
 in formula based on wave equation 42
 experimental determination of 39
Quay, impact on 152–3

Rankine's theory of conjugate stresses 58
Raymond piles 10
Rebound of pile in test 126
Redtenbacher's formula 32
Reinforcement in piles 22, 23, 25
Reliability of driving formulae by statistics 44
Remoulding of clay 79
Resistant bed, location by probing 8
Resonant pile driver 53
Restitution, coefficient of 42
Retaining wall, piled foundation for 161
Rock
 casing tube drilled into 15
 group on 151
 inspection of in borehole 95

INDEX

Rock (*cont.*)
  Oslo point for 92
  pile terminating on 15, 91
  sampling and examination 91
  solution channels in 92
  testing socket for pile 94

Sand
  base resistance 66
  groups in 148–9
  inflow into borehole 15
  soil tests 8
Sander's formula 27
Scale effect, comparison of pile and penetrometer 74
Screw piles 16
Seal at bottom of casing tube 15
Sea water
  concrete in 169
  corrosion of steel piles in 167
Seepage into borehole 14
Sensitive clay
  loss of strength around pile 79
  piles may not be best foundation 147
Settlement
  accuracy of measurement in test 126
  bases in London clay 106–7
  calculated from soil properties 96
  design considerations 96
  empirical rules for 96
  estimation from plate bearing test 94, 108
  group in sand, relation to foundation width 150
  limitation by design 105
  loading test to find 96
  measurement in loading test 124
  of group due to consolidation 144–7
  reduction of shaft resistance with large 108
  single pile not same as group 96
Settlement ratio, definitions 136
Shaft
  continuity of, tested by coring 120
  distribution of shear forces on 62
Shaft compression of bored pile 106
Shaft friction
  movement to mobilise 85
  relation to clay strength 80
  relation to pile material 81
Shaft resistance
  calculation from empirical data 63
  distribution of 43
  experimental determination of 89, 95
  mobilisation with settlement 101–5
Shallow foundation, Terzaghi's solution 61
Shearing of soil at pile base 56
Shear strength, profile of soil 87
Shear zones at pile base 73
Sheet pile 1, 2
Shell pile 10
Ships, impact of 152
Shoe on tube pile 10, 11
Site investigation
  boulders mistaken for rock 8
  by borehole 7
  depth required 7
  influence on choice of foundation 5
  in rock 91
  strength profile of soil 87
  types of 7
Skin friction on shaft 56, 59, 65
  distribution 80, 81
  estimation from Dutch penetrometer 72
  experimental determination of 89
  measured values of 86
Slender piles 109
Slinging piles 23, 24
Soil
  coefficient of lateral reaction of 111
  core, removal of 14
  dilation on pile driving 45
  elastic, pile behaviour in 101
  idealised in early formulae 58
  pressure on shaft 58, 59

Soil (*cont.*)
  properties, ultimate bearing capacity from 56
  samples 7
  strength profile 78, 87
  stresses
    around pile 62, 78
    isobars due to group 135
    release of 13
    relief around borehole 82
    restressing 13
  tests
    in-situ by penetrometer 8
    oedometer 7
    triaxial 7
Soil mechanics applied to piled foundations 56
Solution channels in rock 92
Sonic testing of piles 120–1
Soundness, tests for 120–1
Spacing of piles *see* Groups, spacing
Standard penetration test 7, 68
  corrections to 70
  correlation
    with bearing capacity factors 69
    with density of sand 69
  difficulties below water table 70
  group in sand 149
  relation to pile driving 70, 71
  value taken in design 70
Static formula *see* Chapter 6
  basic assumptions 57
  danger of undue simplification 58
  limitations 57, 58
  soil parameters required for 67
Statistical probability, factor of safety by 97
Steam hammer 17
Steel pile
  corrosion 166–8
  inclinometer duct on 119
Stiff fissured clay, pore pressure change 80
Stress field around pile 10, 63, 90, 91
Stresses in pile due to driving 25
Structural strength of pile 22

Sulphates, damage to concrete by 169–70

Take up 45
  pore pressure dissipation 79–80
Temporary casing 13
Temporary compression 39–41
Tension pile, definition 1
Teredo attack on timber piles 171
Terzaghi, mechanism of pile behaviour 62–3
Testing *see* Loading test
  vibration tests 120
Timber piles
  deterioration of 171
  preservation of 171
  recommended woods 172
  termite attack 172
Tremie concrete 15
Triaxial test for clay 78
  results in fissured clay 87–8
Tube pile 13, 14

Ultimate bearing capacity
  definition of 129
  factors responsible for 56
  formula compared with loading test 43
  by loading test 122, 128
  by static formula *see* Chapter 6
Unconfined compression test for clay 78
Underpinning by pipe piles 14
Underreamed pile *see* Enlarged base
Undisturbed sampling
  of clay 78
  of sand, difficulty of 67
Units of measurement, choice of 3
Upward forces on pile 17

Vibration
  causing soil compaction 7
  methods of soil testing 120

Vibration driving
  basic requirements 51
  combined with jetting 53
  effect of external force 52
  point resistance in 52
  theory of 50
Vibrator driver 17
  amplitude 52
  available models 49
  calculation of exciting force 50
  choice by trial and error 54
  exciting force 50, 52
  experience in Britain 53
  resonant pile driver 53
  for steel H piles 10
  typical Russian pattern 47–8
  withdrawing casing tube 50, 54
Vibratory impact driver 54

Vibro pile 13
Viscous soil resistance 43, 50, 54
Voids in concrete piles 15

Waist in concrete pile 15, 119
Wave of compression in pile 24
Wave equation 33–7, 43
Weak soil below group 150
Weight of hammer 22
Weisbach's formula 43
West piles 10
Wind forces 152

Zone
  of compaction 66
  of shear beneath base 61, 63, 73